Tumble Hitch

Pernille Rørth

Tumble Hitch

A Novel About Life in Science

 Springer

Pernille Rørth
Bisley, Stroud, UK

ISBN 978-3-319-97363-0   ISBN 978-3-319-97364-7   (eBook)
https://doi.org/10.1007/978-3-319-97364-7

Library of Congress Control Number: 2018952465

Cover illustration: Cover figure by © James Milroy (jamesmilroy.co.uk)

This Springer imprint is published by the registered company Springer Nature Switzerland AG
The registered company address is: Gewerbestrasse 11, 6330 Cham, Switzerland

Acknowledgements

I would like to thank Halldor Stefansson and Kai Simons for useful discussions on narrative in science. I would also like to thank Christian Caron, my editor, for his time and effort—and Steve, for everything.

Contents

Part I
The Novel

Tumble Hitch

Chapter 1

"Stop the torture! Stop the torture!"

There was no avoiding them. He kept his eyes on the placard to the far left as it was bouncing up and down with the rhythm of the chant. It showed a kitten. "I feel pain too!" was written below, incongruously. Just as he got close, the cat-suited person holding the placard spun it around to show the all-too-familiar picture: a terrified cat, its head partially shaved and locked in a steel contraption, electrodes sticking out of the skull. He looked away. Pure reflex. The chant went on.

He finally reached the large glass door and gave it a hard pull. It was locked. He looked over his shoulder. The cat-person turned toward him. The furry suit had a smear of red paint across it, but the color was off. The mask was disturbing, though, expressionless but for the flittering eyes in the dark holes. He tried the other door. Same result. The slightest bit of panic crept in, even though he knew it made no sense. One of the security guards was standing right inside the door and spotted him. He pointed to his chest, no, to his dangling access card. Peter understood. He used his own card to open the door and stepped inside.

"Quite a racket out there." He said to the guard. He couldn't see the nametag.

"Been at it since before I got in." The guard shook his heard. "Sir Gerald has been informed. I expect he'll be calling the police. Get it sorted."

Peter nodded and looked back at the small group of protestors—the cats, the rabbits and even a monkey. From where he was standing now, none of them looked menacing. A couple of the suits were incompletely zipped up and held together by safety pins. He could still hear their chant, but it was muffled. He noticed Carol approaching from the outside. She looked fierce and unstoppable. Huifen was right behind her, looking apprehensive. He held up his access card for Carol to see and pointed to the door. She gave a brisk nod and moved toward it, grabbing Huifen's arm on the way.

© Springer Nature Switzerland AG 2018

P. Rørth, *Tumble Hitch*, https://doi.org/10.1007/978-3-319-97364-7_1

What do those idiots think this will accomplish? Peter thought angrily. A few tourists had already started snapping pictures of the spectacle. Yes, Gerald would get it sorted.

—

"Good morning, everyone. Shall we get started?" Gerald entered the conference room with his usual ten-minute delay. "We have quite a bit of ground to cover today and I'm sure everyone is eager to get back to work."

"At least to get away from this steam bath." Hans whispered to Peter. They always sat next to each other at these meetings. With their fair coloring and similar height, people sometimes mistook them for brothers, or at least countrymen. They rarely bothered to correct the mistake. Hans was heavier and had less hair, so Peter's seven extra years were not obvious. Today they sat as far from the sunny windows as they could. The idiosyncrasies of the advanced air-conditioning system in the building had been a popular topic ever since they moved in. Fortunately, most of the thirty-odd people in the room followed the relaxed dress code of academic research. All except Gerald and Bill.

"Will someone open the bloody windows?" Bill growled from the opposite corner. "Please." He added, belatedly. Lucas was up from his seat already. He had come in just before Gerald and had had to take a seat on the sunny side. By the time he had opened the second window, it was clear that it would be temporary. The chanting was loud and clear.

"Stop the torture! Stop the torture!"

It sounded like the same six or eight voices as earlier, with a background of general murmuring, possibly an assembled audience, and traffic. Peter wondered if they were still wearing their costumes. Even in the relative cool of the morning, they had looked hot. He leaned forward to liberate his shirt from the back of the chair and added an answer to seven down on the crossword puzzle discreetly inserted in Hans' notebook. Hans made a "not bad" expression.

"How about we get rid of those raucous prats?" Bill continued. "Gerald, have you considered calling the police?" The tone of casual challenge was not unusual. Why he kept at it, Peter could not understand.

"I have discussed the situation with the police," Gerald responded calmly, "with the chair of our board, and with Sue." He nodded in Sue's direction. "We think the best approach is to let them be, for now. At Moyden, they tried a forceful removal. If you remember, the resulting news-clip of a young female protestor splashed with red paint and manhandled by the police drew public sympathy for the protesters, not for the institute. We'd rather not repeat that here."

"But we don't do any of that here—cats, primates." Akira said. "Do we?" Akira was new in Hans' section and had probably never seen the animal facility.

"No, we don't." Carol answered, quickly and firmly. Peter couldn't see her face. She was sitting not far from Gerald, on their side of the table. "But we do have rabbits. My lab has a couple of projects on early development. And there's the in-house antibody production. Luckily, with the new containment system unauthorized access is pretty much impossible. Silly kids. Bleeding hearts, brains on hold."

"In any case," Gerald continued, "if you do pass by these spirited young people when you go out, please be polite. Or say nothing. No arguing, please. No provoking them. They don't seem to have found the parking garage entrance, so use that if you prefer. Now, Lucas, if you could—"he nodded at the two open windows"—and I will try to make this meeting short."

There was a general mumbling of assent as the windows were closed.

"As you all know, the written material for the institute review in November is due by the end of the day today." Gerald continued. "Please hand your complete dossier to Shirley before five p.m., if you haven't done so already: C.V., the write-up of research accomplishments and plans for your group, with three selected reprints."

"Why so far in advance?" Darya said, with some irritation. "I have two papers still under review."

"That's just how it's done. But don't worry—you can update your C.V. later on." Gerald explained how.

Peter smiled to himself. He had already given Darya feedback on her write-up. She had nothing to worry about, whether her latest work was in press or not. But he knew her final version would not be handed in until six p.m., when Shirley packed up for the day. The same for Lucas. Lucas had also asked Carol for feedback, Peter knew. Smart. She understood the mouse aspect better than he did. For both Darya and Lucas, it was their first review. They had a lot to prove.

———

He took the garage exit on the way out, but returned the usual way from the sandwich shop. Simple force of habit. Once he was close enough to see them, it also seemed silly to change course. Some of the animal suits had been abandoned and half of the masks were off. The chant started up again as he approached. A young man broke from the group and moved rapidly at an angle to Peter, as if determined to intercept him before he reached the building. A girl, or a young woman, noticed the move and followed him. Peter did not change his stride and tried his best to maintain a friendly smile. He knew his height could be intimidating to some and this man was on the slight side. As they got closer, Peter noticed that he was not actually that young, early thirties perhaps. His expression matched the intensity of his approach. The girl following him was twenty or so, about the same height as the man and had blond hair in a ponytail. She walked fast and had almost caught up by the time the man reached Peter. She looked apprehensive.

"We know what you do in there. Torturing innocent animals. You should be ashamed of yourselves."

"We don't—" Peter started, then stopped. He also stopped walking. The man stopped a few feet away, the girl next to him. The man had the look of someone hoping to be provoked. Peter swallowed his words, reset his smile and held out his hand.

"Peter Dahl. I work at the Codon Institute, as you may have guessed. I study memory in fruit flies."

"Fruit flies? What the bleeding ..."

"Tina." The girl interrupted him, and shook Peter's hand. "And this is Alistair." Alistair shook hands as well, stumped for the moment. "We are not protesting science in general," the girl continued rapidly, "but we know that some scientists at your institute do experiments with rabbits and mice, mistreating them and their babies. It's cruel, inhumane. . ."

"You deliberately induce pain and suffering, torturing the weak and the vulnerable." Alistair had his voice back and took a step forward. "All in the name of 'research'. Because they are not of the ruling Homo Sapiens species, you accept the unacceptable. Torture. It's the Holocaust all over again."

It was Peter's turn to be momentarily speechless. He had not expected the Holocaust. It was too outrageous. "That's. . ." For lack of a good answer, he turned and walked briskly to the door, card already in hand. When he looked back, Alistair and Tina were still standing where he had left them. Alistair's expression was triumphant. The girl, Tina, looked puzzled, as if she was trying to work something out. Peter turned his back and headed for the stairs.

——

"Is this for real?"

Peter was standing behind Mihai's fly-station. The light was on, but no flies were on his pad. Mihai swirled around to face him. He cocked his head with an expression of mild amusement. He often took his time in answering. "The graph you left on my desk." Peter continued, holding it up. "Are you sure you haven't swapped them?"

Mihai smiled. "Of course I haven't. The wild type curve decays as normal. The mutant just does much better."

"Yeah, I see that now. But it almost looks too good to be true." Peter looked at the graph again and tried to hold back a big grin. "So. What's the mutant?"

"I'll let you know when I've outcrossed and retested it."

"Oh, come on. . ."

Mihai shook his head. His hair was very short, probably shaved. Combined with a body that appeared shaped by serious weightlifting, the gentleness of the face, and indeed of Mihai himself, was unexpected. He was mature for a PhD student, hardworking and ambitious. Today he would not budge.

Peter looked at the piece of paper again. "BD, zero, zero, one, one? Really? And just the one fly?"

"Yes, but he's been mated and there are progeny coming. I've done the PCR, as well. It looks—interesting. But I won't know for sure until I've tested outcrossed offspring." His smile seemed to contradict his statement. "I will also order and test RNAi lines. Then I will tell you which gene it is."

"You can't even give me a hint?"

"Delayed gratification builds character, you know."

Peter looked at Julia, who was sitting at the next fly station over, busy with her own work.

"Don't look at me." She glanced up. "He won't tell me either." She smiled, briefly. She and Mihai got along well.

Peter emitted a low groan and shook his head. This was an exquisite kind of torture. But he would let Mihai tease him and let him savor the result on his own for a while. He had earned it. Peter had given advice and ideas, but Mihai had done all of the hard work. He'd built the tracking apparatus with a fellow student from engineering, designed the "bar-code" system and spent many months in the fly-room screening the large array of mutants. For Mihai, this would be his first big result.

"So a month, two?"

"More. Another three generations, and then the behavioral tests. I need a really long time curve to check their memory properly."

Peter threw up his hands in mock-exasperation.

"Someone seems to need to talk to you." Mihai nodded toward the doorway, where Lucas was waiting, looking anxious.

"I'll be right there, Lucas. Just take a seat in my office."

Peter watched the pink shirt and the longish curly black hair retreat from the door. When he turned back around, Mihai had already picked up a fly vial and turned on the carbon dioxide. Conversation over for now, it seemed. Peter resisted the urge to plant a congratulatory pat on the massive shoulders. He turned to Julia instead.

"Can we talk to tomorrow morning? This afternoon is a bit crazy."

"Sure. Whenever. I'm in early." Julia said, not taking her eyes from the eyepieces. "I don't have much new to show you, though."

"Still, let's have a look, shall we? I'm curious about the override experiment. It looked promising." She didn't turn but a smile appeared on her face. She was just starting out, but he knew she was both smart and eager. She would do well, he thought.

———

For someone with a deadline in a few hours, Lucas seemed strangely reluctant to finish their conversation. Or maybe not so strange, Peter mused, he probably needed reassurance more than the document needed editing. Peter did the best he could. He was finally rescued by Ilana at the door. She had her laptop and notepad in hand.

"Come in, Ilana, come in." Peter stood as he waved his hands loosely about, introducing. "Lucas, you've met my postdoc Ilana, haven't you?"

"Yes." Lucas got up, but didn't smile or look directly at Ilana. Instead, he headed straight for the door.

A bit rude, Peter thought, but perhaps he was too preoccupied. "My door is always open." He said to the receding pink shirt. "Anytime."

Peter felt a sudden relief that he had suggested for Ilana to collaborate with Carol's lab rather than Lucas'. Lucas probably needed to get more settled before he'd have time for a run-of-the-mill collaboration. Ilana was focused and eager to embrace the mouse model. Luckily, Carol and her postdoc Huifen had volunteered to help. Peter was well aware how important the project was his lab, even if it was unlikely to provide the novelty and excitement that he so loved from other projects.

"I need a data-fix." He said cheerily as Ilana got seated. She looked puzzled, so he added. "I mean, I hope you have some results for us to discuss. I love fresh results." He smiled again, but then let it be. Ilana never joked or laughed, at least not around him. Combined with the straight, black hair, the angular face and dark-rimmed eyes, her somber demeanor suggested a dour seriousness. But Peter felt it was more likely just insecurity. Despite her seniority, she had none of the natural confidence of someone like Mihai. Peter knew he probably couldn't do anything about that.

"I have done the first passive avoidance tests." She answered, her voice now behind him. He had stepped across the room to open the office door that Lucas had closed on his way out. He had never done anything the least bit inappropriate with his female students and postdocs, but they had been told to be careful about closed doors. "I'm still trying to get the parameters right." She continued. "I just want to be sure..."

Peter sat back down and turned his full attention to the graph displayed on Ilana's laptop. They talked about passive and active avoidance tests and about the logistics of genotypes, age and time courses for well over an hour.

When they were done, Ilana also seemed reluctant to leave the office. Her reason was different from Lucas', though.

"Is it safe to go to the animal facility?" She asked. "With the protestors around? They are a bit scary."

"Don't worry. They can't get inside—and certainly not into the facility. The security system is good." He paused. "But we may have to put up with them for a while." She nodded, but did not look reassured.

—

"You have a visitor. She's in reception."

"Can you just send her up?" He looked at the time: it was close to five. He had no memory of an appointment, but it wouldn't be the first time he had forgotten.

"No can do. No ID, so I can't issue a pass."

"OK. I'll be right down."

She was alone in the reception area, no Alistair in sight.

"I'm Tina." She reintroduced herself. "We met outside earlier."

"Ye-es?" She had to know that didn't exactly recommend her. "How may I help you?" He continued, his tone clipped.

"After we met, I looked you up. I—I mean we—we understand that you do important work on how the brain works, on memory." Her voice was hesitant. She kept looking down. "And that you don't use—you don't hurt—real animals."

"You could say that. People usually aren't too concerned about us swatting a few flies." She seemed to twitch a bit at this, or maybe that was just his imagination.

"I was wondering—I mean—we were wondering, if you would be willing to give an interview about your work? For our newsletter? News of Eden?" She paused. "Our group is called Eden." She drew a breath. "We would like to show that that there are alternatives to cruelty—I mean—that the three Rs can really work." She finally looked bit more confident.

So this girl knew about Replacement, Reduction and Refinement? Peter thought. Was she the moderate wing of the little group? Well, it was a far cry from invoking the Holocaust. At least she was trying. There was also something in her voice that he recognized. The accent.

"Are you Danish?" He asked, in Danish.

"Yes." She answered, but in English. She looked down again for a moment. Maybe she did not like her accent being detected. "But the newsletter is in English, of course. So I prefer if we speak English."

"Sure," he said, smiling despite himself, "but I'll have to ask our director first." Gerald had a good feel for how best to interact with the public. Peter actually wanted his advice. "Anyway, I don't have time this afternoon."

"Tomorrow. Can I come back tomorrow?"

"Come around three p.m. and we'll see, OK?"

"OK. Thank you."

She seemed oddly relieved, with the delay or with the possibility of success, he could not tell. She added the briefest of smiles and hurried back outside. He stood for a moment, watching her back as she moved toward the remains of the group. The ponytail swung back and forth like a quickened pendulum.

* * *

"I'm out back." Jessie yelled, sounding cheerful.

She must have heard the front door slam shut. The sudden draft when he opened it had taken him by surprise. He was usually home first, not having quite as many last-minute emergencies as she did.

"I'll be right down." He said, loud enough for her to hear. "I just have to wash my face. I'm all. . . " The last bit was more of a mumble.

With a partially dried face and his shirt flapping, he took the steps to the lower ground two at a time. He passed the dining area, deposited the bags on the kitchen counter and looked into their small courtyard while doing up a couple of his shirt buttons. The rear wall beyond the kitchen area was almost all glass: two large windows and a sliding glass door, which was currently open. Jessie was sitting in her favorite fair-weather spot, the wooden patio swing, facing outwards. The swing was moving gently and the big outdoor fan appeared to be blowing straight at her. Her auburn hair was just long enough to dance in the breeze. She turned her head and greeted him with a generous smile. The smile made her face glow. It never failed to cheer him and to remind him of his luck.

"In the fridge." She said, holding up a glass in one hand. It was a tall glass and had something green in it. Maybe mint leaves and lime with sparkling water, one of their recent favorites. Or, even better, a Mojito. He poured one for himself from the narrow glass beaker in the fridge door and tasted it. Yes, he thought, and smiled, a perfect Mojito. He stepped out onto the flagstone-covered patio. Over the years, they had found plants tolerant of the courtyard's near-constant shade and pampered each with a generous pot. Some moved inside for the winter. The plants covered most of the rear and side walls and gave the place an almost tropical feel.

He walked up behind the now stilled swing and bent down, folding his long arms around her, his cold glass held to one side. She leaned back for a moment, clasped his arms in place and gave them a friendly squeeze. Then she turned her head, lifted up and planted a kiss on his mouth, another on his cheek and ruffled his slightly wet, slightly blond and slightly gray hair. The swing moved below them. He smiled and kissed her back, then unwrapped from the embrace and stepped over to place his face in front of the fan.

"Crazy weather." He said and lowered himself into his wicker chair. It creaked and groaned but did not move. He would never get used to the swing.

"I suppose we shouldn't complain when we finally have a bit of real summer, should we?" Jessie took a sip from her near-empty glass and pushed off to get more movement from the swing. "But it is a bit much. I almost died in the office." She looked very content, nevertheless. "At least we have ceiling fans and our old-fashioned windows can be opened. That helped. The traffic-noise made conversation almost impossible, though." She shook her head, still smiling. "So how did it go today? Did everyone get their write-ups in on time?"

"The usual last-minute panics, of course. I know Shirley won't send it off until tomorrow or the day after. Gerald always wants to give it a last check. But moving the deadline wouldn't help."

"It's human nature, I suppose." She paused a moment. "You've read the ones from your section, haven't you? Will they do alright?"

"Yes. I'm quite sure they will. Anyway, the panel tends to cut young group leaders some slack."

"As they should."

"As they should. I'm happy I don't have any of the old guys from the Lester Institute in my section. Their science gets reviewed along with everyone else's, even though, well... You know..." He shrugged. She nodded. "Awkward. But they have tenure, so we have to live with it for a few more years." He paused. "Overall, I think the Institute is in very good shape. Gerald is doing a great job."

"Hmm." With a thoughtful expression, she tilted her head, stopped the swing and put a hand on his knee. "You know..." she started, but stopped again when he didn't look up immediately.

"A strange thing happened today." He said, picking up her hand and studying it. "There was a group of protestors at the institute, out front. Animal rights stuff. They were wearing animal suits and masks and waving placards... and chanting... "Stop the torture". It was quite unpleasant."

"The Codon Institute is not an obvious target for that kind of thing, is it? You don't do any of the contentious stuff, do you?"

"No. We don't have any primates, and no cats or dogs."

"I remember a few years ago, when activists were in the news a lot. But I thought they had cracked down hard on them after the harassment campaigns."

"They did. The ones I saw today seem like amateurs by comparison. Luckily, our animal facility is out of sight, in a connected building. Anyway—the strange thing was not so much them being there, but that one of them asked to talk to me. Me, specifically. She says they want to interview me because they like my doing brain research without cruelty to, quote "real animals" unquote."

"That's kind of nice, isn't it? It suggests more awareness of what research is than you normally get from that front."

"Yes, I suppose so. Anyway, this girl turned out to be Danish. Not the rest of the group, I'm pretty sure, just her."

"And you think that's why she asked to talk to you?"

"I don't know. I guess it doesn't really matter. I asked Gerald and he thinks it's a good idea. We should do whatever we can to help resolve it peacefully, he said. Maybe it'll make them go elsewhere. So I'll be giving the interview tomorrow. I hope it doesn't prove to be a mistake."

"How could it be?" She asked but got no response. He seemed far away. Grabbing one of the side poles of the swing, she pulled herself upright. "Refill?" She asked, now standing right in front of him. "There should be another glass in there. Then we can make the second ice cream for tomorrow. Maybe before we eat? Before we collapse?"

"Right, tomorrow..." His voice trailed away. Finally, he seemed to snap to, drained his glass quickly and handed it to her. "It's great, the Mojito. Exactly what I needed. Oh, and I left the take-out on the counter." He stopped and looked at her face quizzically. "Ice cream?"

"Tomorrow evening. Dinner. Hans and Alessandra, Nigel and Beatrice. We just have to do dessert. I got some berries."

"Oh yes, of course. Remind me, why Nigel and Beatrice? Why not Carol and Wi...—Carol and Not-Bill? They are much more fun."

"We owe Nigel and Beatrice one. And I think William is touring right now." She smiled, a bit mischievously. "We really should stop with the Not-Bill."

"He can take it, I'm sure. After all, he's been married to Carol forever."

"Right. But he'll never bring his violin along if you keep teasing him. I still have this dream of getting a private recital some day."

"Me?" He laughed. "You started it. Not-Will, then Not-Bill. It was definitely you, teasing the stiff Brit."

Jessie smiled and shrugged before turning around to go inside. That was when Peter finally remembered.

"So what was that thing you wanted to tell me about?" He said, raising his voice to be heard. She was in the kitchen already. "Your home-going email sounded tantalizing, but it was very vague."

"I thought it better that way." She said when she returned. "My work email isn't private." She handed him an almost-full glass. "Plus, I wanted to tell you in person." She sat down on the edge of the swing, keeping it steady. Then she put her drink down on the small metal table and folded her hands in front of her knees.

"Tony called." She finally said.

"He called? I thought he was always too busy to call."

"He is. But this was special." She paused. He understood her pauses. She was looking for the best words, for the optimal approach. But she was smiling. "You know about Oak Hill?"

"Sure. The new brain institute he's setting up. It sounds interesting—and very ambitious."

"It's there. Construction has been completed, the labs are almost ready and they are starting to hire."

"They are?" Peter's chair creaked as he straightened up.

"Yes. They are. And you won't believe this." She stopped again. His curiosity made him lean further forward. She picked up one of his hands in both of hers and continued. "He asked me if I might be interested in being deputy director. Deputy director in charge of scientific strategy and communication." She beamed and squeezed his hand hard.

Peter kept his anticipatory smile for a few seconds more. It wavered a bit, but then he brought it back, redoubled.

"That's fantastic. Deputy director. At Oak Hill. Wow. I would have—no, forget that. I'm so proud of you. That's amazing."

"Right," she said, "I know what you were thinking. I'd have thought he would choose an active scientist too, not a mere scientific editor like me." She made a dismissive gesture.

"Editor-in-chief of the most influential journal in our area. No unnecessary modesty here, OK?"

"Well, anyway. I guess he sees my experience as relevant."

"I'd think so. It's an inspired choice, actually. You are well known, well respected, and you know the whole field better than any of us narrow-minded practitioners."

"I suppose. . ." She smiled and paused. "But that's not main reason, actually. Do you remember this crazy idea I came up with last year? Right after I visited the Franklin Institute?" He shook his head, furrowing his brow. She continued. "It was about a new way of publishing science, based on unique, insightful observations followed by comments. I'm sure I told you about it. I was a bit obsessed with it for a while."

"The journal wasn't interested." He nodded, vaguely remembering how excited she'd been at the time.

"No, they weren't. That's not so surprising, really. The current model works very well for the publishing houses. No need to rock the boat." She paused. "Anyway. I also talked to Tony about the idea back then. We met at a conference in the US not long after my attempt to convince the journal. Apparently, he hasn't forgotten about it. He's considering taking up the idea, establishing it at Oak Hill." Her eyes widened. "Obviously, the actual science being done at the institute will be his major focus. But he has, he says, a broader vision of what Oak Hill should be contributing." She couldn't stop a big grin from appearing.

"Which includes launching a completely new initiative in science publishing." He filled in, dutifully.

"Well, at least trying it out." She nodded. "You remember, part of the uncertainty I had initially was whether there had to be a physical host institution."

"Yes." He didn't really remember the details, but also didn't feel like asking.

"So." She said, with emphasis. "That's why he wants me, specifically, for this job. So we can give the idea a chance. The job would probably involve lots of other things as well. We haven't gone into details about all that."

"That's—that's really—something." He said, quietly.

"Oak Hill would be a perfect fit for you, as well." She added, hurriedly. "The place will be full of people interested in brain function and memory-formation. A state-of-the-art place... New, exciting, special." She drew a deep breath. "If you think you might be willing to. . .?"

"Of course." He smiled and looked steadily at her slightly flushed face. "It sounds like a fantastic opportunity. You can't not go for it."

"Well, for now, I'm just going to go see the place and talk some more with Tony. He knew that I'd be in the US the next couple of weeks and asked me to come by toward the end of the trip." This time, she took both of Peter's large hands in her own. "So you're sure you're good with this? You would consider moving back to the US?"

"Yes, of course. We moved here for my job, primarily. Now it's your turn."

"Well, London was a good move for me as well, given the journals that are based here."

"Still. I got exactly what I wanted back then. It's important you get a chance to do exactly what you want. I support you one hundred percent."

"Well." Jessie was smiling so much, it looked like she would burst. "I suppose I'll know more when I've been to visit. But it sounds so. . ."

"Fantastic. It does. Anyway, maybe a move would be good for me. I've been at the Codon for what, eighteen years? It's been great, but. . . It's a long time. And at fifty-two, I probably only have one serious move left in me. From what I hear, Oak Hill will be a fantastic place. *The* place for brain research."

"It will be, I'm sure."

She kept smiling. Peter liberated his hands and rubbed her knees.

"So did Tony. . . did he mention me, directly?"

"We didn't get that far. I couldn't. . ."

"No, of course not." He added quickly. "That would have been premature. We'll know more once you've been for a visit."

"I know Tony really likes your work. He's told me so on several occasions. I mean, didn't he introduce you as "the man who brought genetics..."

"The man who brought the power of genetics to the enigma of memory." He laughed, softly. "Yes, he did. Very flattering. . . At least he didn't call me the father of the field. Then I would have felt really old."

"Tony is older than you are."

"I know—just joking." He got up from the protesting chair and headed for the kitchen. "This calls for bubbles, don't you think?"

Chapter 2

Bat out of Hell gave way to incantations of wasted youth. The relentless beat and the raw energy of his voice helped, but Jessie still felt sluggish. When she got up, at her usual early hour, she had felt ready to tackle the world. Now yesterday's celebratory bottle and the not-so-early night were taking their toll. She looked at the display on the elliptical. Still a ways to go. She had waved at Rissa, also on her usual schedule, ten minutes ago. The thought of telling Rissa her news gave Jessie some extra energy, much needed.

On the screen, four minimally clad and well-oiled bodies gyrated in provocative parallel routines. The mismatch to the music in her ears might be making it worse, but the in-your-face faux sex made her feel uncomfortable. She looked away. Up front, two fully clad women were walking leisurely on the treadmills. They seemed unperturbed. Her gaze moved on to the new regular, whoever she was, always here, and ended on the painfully skinny girl at the end of the row. It was hard to look at her, hard not to. Jessie had felt tempted to talk to her several times. But what could she say? She knew it wouldn't help. She looked at her display again. Almost done. Five miles on the treadmill, then stretching, then she'd catch up with Rissa.

———

"He's too good to be true."

Rissa shook her head while taking the first careful sip of her banana smoothie. Jessie had given her the short version of yesterday's events between the showers, the locker room and the juice bar. They always stopped at the bar for a smoothie and a chat. Rissa nodded, smiled at the beach-boy type behind the bar and touched her card to the reader. "And I've still never seen him." She continued. "I'm beginning to think you made him up, this husband of yours."

"For twenty years?" Jessie smiled. "That's one long-lasting fantasy. Anyway, you have met Peter."

"I have?"

"He came to an opening at the gallery a few years ago, when you had just moved to the new space."

"Oh, that one." Rissa lifted an eyebrow.

"I know. You've had more than a few openings. I don't remember which one it was. Or whether he liked it. But you have met him; he *is* real. He's just more supportive than most husbands."

"Well, you've earned it, girl." Rissa flashed a wide smile, her perfect white teeth shining. "Now this," she added, pointing to the yellowish mush with her extra-wide straw, "is the real reason I come here every morning."

Jessie sent her a look.

"OK, so not *every* morning." Rissa said. "Too much socializing in my line of work for that. It's so tiring."

"But I thought you enjoyed that part."

"I do. Of course I do. I'm just getting older—" she widened her deep brown eyes, not yet decorated for the day "—practically middle-aged." She added dramatically.

"That's not a word I would ever associate with you."

"And don't you dare!" Rissa smiled. "The tragedy is, I still haven't managed to make it to the gym as early as you do. Not once. I've never seen the muscle-men you claim are here from the small hours."

"I'm pretty sure they are bouncers, beefing up on their way home from work at the clubs. You could come in with them. Beat me to it."

"I might be able to motivate myself some day, some night—just to see them. I suspect you've made them up as well."

"You give me too much credit." Jessie shook her head. "I suppose you have seen skin-and-bones, though, the scary one. And that new girl."

"I haven't noticed. I have my priorities."

"Always looking. . ."

"Well, and?" It was a long-standing mystery to Jessie. How could someone like Rissa be single? She was gorgeous, successful and fun to be with. She had originally come over with someone, way back. But he had not stayed. Others had been interested, but none had lasted longer than a few months. The break-ups had been tough. This was how they got to know each other better. Jessie was a good listener. At least Rissa seemed to think so.

"Anyway, I don't even know when you start." Rissa continued, bringing Jessie back to the present. "In the morning."

Jessie smiled. "Too early for you, girl."

Rissa laughed. "Don't you "girl" me. You can't pull that off. No chance."

Jessie held out her hands, palms up, then palms down, looking at them with a mock-sad expression.

"That's not your only problem, girl. You've almost lost your twang. Maybe going back home is what you need. Connect with your roots and all that. Before you get terminally corrupted by the cosmopolitan Londoners."

"To be honest, I'm not sure where home is anymore. I like it here. Better, I think. I fit in. But this job is a great opportunity."

"Good for you." Rissa said, emphatically. She continued in a softer tone. "I will miss you if you go, you know? My anchor of sanity and reality when my" she added finger quotes, "temperamental artists are driving me up the walls."

"What a recommendation: sane and real." Jessie smiled, nevertheless.

"You know what I mean. All these years." Rissa reached out and rubbed Jessie's arm very briefly. That was unusual for Rissa—physical contact.

"I do know what you mean." Jessie found tears threatening. It was ridiculous, she told herself. Get a grip. The tears receded. "But this means my trip has been extended for three more days. I'm afraid I'll miss the opening of your new show. What was it? Something lines?"

"Fine lines and big form. Drawings, paintings and sculpture by female artists. It's quite powerful stuff, some of it. When will you be coming back?"

"July seventh."

"No problem then. The opening is on the twelfth. I'm afraid you have no excuses." Rissa looked at her with intent. "I need you there."

"Your sanity and what-not."

"My assurance that I make it home afterwards. In fact, that proves my point. Your man is too good to be true. He lets you stay out with me until the small hours and he never doubts you."

"True. Maybe I should be worried."

"Nope. You're just lucky." Rissa smiled while she shook her head. She glanced at the oversized, bright pink and green watch on her wrist. "Time for you to hit the road, no?"

"Right. I still have a regular job to do."

"Yes, whatever it is. I prefer not to know."

Jessie took another long sip of her smoothie. "What do you do in before nine anyway? Isn't everyone in your world still asleep?"

"I cook the books, of course. No, seriously, people with money don't sleep late. My buyers. You forget that I'm a business-woman." Rissa gave a coquettish shake of her shoulders, touched her handbag and crossed her expensively-shoed legs. They had both done very well, professionally, since they first met.

This would be the hardest part of leaving, Jessie thought, losing this. She picked up her gym bag and finished off her strawberry smoothie.

"See you in a couple of weeks." Rissa said, waving her off. "Break a leg. And say Hi to that charming brother of yours, will you?"

"Ah—him you remember." Jessie said. That was years ago, she thought, his visit to London. Eight years, ten? "He's still married, you know."

"All the good ones are."

———

"Jessie Aitkin speaking. And yes, I am the idiot in charge, as you so nicely put it to one of my editors last night."

Normally, Jessie wouldn't take these calls. Most authors seem to calm down when they have to put their rebuttal into writing. They are more focused and repeat themselves less. And they don't yell. But Charles Baker was a special case. He was a bully, and even more belligerent in writing. He seemed to wind himself up in his eight to ten page tirades, rather than cooling down. And, more often than not, a second missive would arrive before the first one had even been dealt with. So she had decided to try a conversation. But she had been dreading it.

"I used to have the utmost respect for this journal." He started, the voice quivering with indignation. "I have sent my very best work to you people. Time and time again." The standard seniority intro, she thought, I can guess what is coming next. After a short pause, Baker continued. "So I am appalled by this treatment of my most recent manuscript. Rejected, by a junior editor. A nobody, who has never made a single contribution to science." He paused again, allowing her to protest. She did not. "Have you even read the paper yourself?" Jessie said that she had, as well as the reviewers' comments. Saniya, the unhappy junior handling the paper had alerted her to the case as soon as she got in this morning. The rejection was a straightforward one, based on three reviewers' evaluations and recommendations. "This is very important work." Baker continued. "I know when something is significant and when it is not." Except if it is your own work, Jessie thought. "I mean, when I see what you publish. Only last month. . ." He went off into a rant about someone else's inferior paper before getting back on course. "If you do not reverse this decision, I will stop sending you papers altogether. I will make sure everyone in the field knows about the level of incompetence in your office."

"You have concerns about two of the reviewers." Jessie stated in place of a direct response.

"Yes." He stopped and re-started awkwardly. "Now, the first reviewer is very positive. He only has minor corrections." But he actually did not rate the paper very highly, Jessie knew. Competent work, he had said, in the "may be published" category, not the "should be published" category.

"The second reviewer is obviously on some kind of personal vendetta and does not appreciate the importance of mouse work. He should be disqualified." She, Jessie thought, not he. She just happens to be our foremost expert in the area and is usually a very fair reviewer. Her main point, about lack of conceptual novelty, is well argued.

"Now the third reviewer is both incompetent and unreasonably picky. The review is full of mistakes. And the experiments he asks for are completely unnecessary." Baker went on for a while about reviewer number three. Jessie jotted down some notes as he spoke. Unfortunately, he was right about quite a bit of this. Perhaps the reviewing had been done hastily, or done by someone in the lab who didn't know as much as he or she thought. Shit. There was nothing worse than mistakes like this for an author to obsess about. The fact that none of the reviewers had said the work was important enough to be published in the Journal was disappearing into the thick fog of details and arguments. This would take hours and hours to sort out.

"Now this editor of yours, well, I am shocked that someone with so little knowledge of the field could be given the responsibility to make decisions about important papers. It is clear that she does not grasp the importance of. . ."

"Dr. Baker. Let me suggest we stop there. I will take another look and make a decision. Not because I mistrust the judgment of the primary editor, or the reviewers, who actually all agree that the paper is not for the Journal, but because I appreciate that it is important to be thorough. But please, no threats and no deriding my colleagues. It does not help."

"Exactly. It's all just a job to you . . ."

"Goodbye, Dr. Baker. I will be in touch via email."

Jessie's left ear was warm and irritated when she finally hung up. The ceiling fan was churning, cars blowing their horns impatiently and her inbox endlessly growing. She still had the rest of the day to get through.

* * *

They had stepped up the campaign. This morning, they also had two very real-looking metal cages. Meant for dogs, perhaps? They were certainly too small for an adult human—the caged and costumed protestors looked very uncomfortable. But the trick worked. It was hard to look away. Some passer-byes walked faster, others stopped up. More pictures were taken on more phones.

For lunch, Peter opted for a sandwich from the in-house stall. The bread was so stale he threw the last bit in the bin.

—

"I will sign her in," he said to the skeptical security guard, "she's my visitor." No ID, no last name even. Just Tina. He was as irritated at the lack of co-operation as the guard was, but endeavored not to show it. Gerald wanted to try the accommodating approach and Peter was the first one on the line. He smiled briefly and looked at her sideways while filling in the form. She did not look the least bit threatening, her face pale and uncertain, her gray eyes flickering, the ponytail emphasizing her youth. She wore a loose skirt and blouse, both well worn, and carried a battered laptop sticking out of a woven shoulder-bag. They exchanged only a few words as they ascended the central stairs. Unfortunately, she accepted his routine offer of coffee from their espresso machine. This meant he was distracted for a few minutes in the common area. She stayed close, though, and looked at a newspaper left behind by someone. No one approached them or asked who she was. She could easily be an undergraduate looking for a summer placement, or a starting PhD student. Everything looked perfectly normal.

"We go down that corridor, third door on the left." He tried to indicate but both his hands held coffee. She seemed to understand and opened the correct door. They went in. He left the door open.

"So, where should I start? I'm not sure how much science background your readers have. You want it in English, right?"

"Yes, please. The Eden group is. . . We have different backgrounds. We all love animals but. . ." She busied herself answering the laptop's call for attention. Then she looked up. "I have a list of questions. For the article."

Her eyes stayed on his face for a bit longer than normal, he thought, and this gave way to a hesitant smile before she focused on the laptop screen again. More curiosity than hostility, he sensed.

"Sure. That sounds fine." He said. "Fire away."

"OK, first question then: You study memory in flies."

"Yes. . ." He said, not sure if this was a question. "Fruit flies, Drosophila."

"These fruit flies," she continued. "they are insects, and they are tiny. Yet you claim we can learn about human brains by studying them. How is that possible?"

"That's a bit tricky to answer." He paused. "Obviously our brain is much, much larger. But the basic units of it, the cells that make up the brain, are quite similar. They have the same kinds of gated channels" she looked puzzled "sodium, potassium and calcium channels that can be controlled by voltage or by. . ." That didn't help. He sighed. The same kind of molecules would be too vague. "These cells," he tried again, "they talk to each other in the same way. Fly brain cells and our brain cells. More or less." He shook his head. "Can we return to this question later?" He smiled, almost apologetically. She flashed the hesitant smile from earlier. Funny, it already seemed familiar. She looked at her screen again.

"How do you know that flies have memories? That they remember?"

"A good question." Simpler to answer, he thought, with relief. "Well, we teach the flies something new, test them right away to confirm that they did, in fact, learn. Then we test again later on to see whether they remember. Some do, some don't. Over time, they forget."

Reassuringly, she nodded. "Can you give me a specific example?"

"Sure. Say we expose the flies to a neutral odor. That is, something they can smell but that doesn't automatically mean food or danger or a possible mate for them." She nodded again. "A moment later we give them an electric shock." He noticed a reaction in her face, quickly suppressed. Wrong choice. He hurried on. "After a few repeats of this, they avoid the odor, given a choice." He drew a quick breath and continued. "It also works the other way around. We can train flies to associate something they like, say sugar, with a neutral odor. Afterwards, they will tend to walk or fly toward this odor, looking for sugar." He smiled, slightly. "Untrained flies won't. So, obviously these flies have learned to associate the odor with the sugar reward during training. It is not something hardwired in their brains by evolution. We then measure how long they remember the positive association. With simple experiments like these, we can determine what genes affect the ability of flies to learn, and their memory. Flies have long-term memory and short term memory, just like we do."

"Do they feel pain?"

He should have known she would pick up on that. He tried to explain that it was hard to tell. The flies avoided shocks, just like they avoided too much heat or cold. It was hard to know if they experienced it as pain, as such. He was going in circles. He hoped for a better question.

"What do you learn, when you study memory in flies?"

"Big question." She was about to say more, but he moved ahead quickly. "But I do have some answers. Small answers, I suppose. We've identified a number of genes that are necessary for making memories. Let's say that a fly doesn't have enough of one of the gene products or that it's defective. When we test it, we find that this fly can still smell and move normally. It may even be able to learn to associate sugar with the odor. But if we test it a few hours later, it has forgotten." He paused briefly, smiled. "And sometimes—and this is the best—we can tinker with the genes so that the flies learn more effectively or remember for longer. That's rare, though. But if we can show that having less of some molecule improves memory, while more makes it worse, for example, then we can be fairly sure we are onto something significant."

"So, you are saying," she started, furrowing her brow, "that simply reducing the amount of something the fly has in its brain can make it better at remembering?"

"Mostly not, but occasionally yes."

"And the same might be true for people?"

"Yes. Well, we think so, but we haven't. . ."

"So that means flies are not as good at remembering things as they might easily be? That they are designed to forget? That we are?"

"Well, in a way, yes. If by design, you mean evolution. Brains have evolved to deal with nature as the animal experiences it. It may not be beneficial to make an association too easily, or to be able remember it forever."

"Co-occurrences could be by chance, random." She said quickly, tilting her head with an expression in between delight and curiosity. "Some associations might not be worth paying attention to, or remembering."

She is smart, he realized with a small shock.

"Conditions may change, old associations may become irrelevant." He added, eagerly. "Most animals have evolved to adjust. Of course, humans may forget things for more complicated reasons, as well."

"Yes, of course." She said, the sudden eagerness gone. Her eyes left his face and settled on the screen again. A moment later she continued in a neutral tone, apparently reading off the screen. "What made you choose to use flies for your research?"

He flashed a grin and started talking about mutant flies and genetic screens, generation time and sequenced genomes, conserved pathways and homologies. He realized that she was unlikely to understand or retain all of it, but he found it strangely difficult to stop.

"What is the most important thing you have discovered?"

"Thing? Do you mean an abstract principle? Or the function of a specific gene? We have found several of those."

"Genes, then."

He told her about the best ones, with a bit of pride: What they were needed for and how his lab had found them. Numbskull, for example, and Dummkopf. She seemed to find the nomenclature amusing. He explained how they named them. While doing so, he noticed that he was leaning forward in his chair. He pulled back, discretely. She glanced at her screen and furrowed her brow again.

"Is it possible to test drugs in flies?" She asked.

"Absolutely. A number of labs are testing potential drugs on fly models of cancer, or even dementia. It's quite efficient and much more physiological than, say, cell culture. So—"

"So," she interrupted, her expression suddenly combative, "you're saying it isn't actually necessary to do all those experiments with rabbits, cats and monkeys?"

He pulled further back in his chair and took a deep breath. He reminded himself that this was not just a friendly chat. With care, caution and a good measure of reluctance, he explained about the limitations of work done in insects and other non-mammals. Not everything was transferable to humans—and never directly. Slightly different genes, differences in physiology, drug metabolism, side effects. . . He knew he had to say all this.

They were quiet for a while afterwards. He glanced at her briefly. What was she thinking? "So, do you have enough for your article?" He asked. She didn't respond. "If you have more questions, I'll give it a try. Just not questions about furry animals, OK? It really isn't my area." He said the last part softly, cautiously.

It took a while before she spoke again, and when she did, her expression had changed. She looked inquisitive again, but with an extra layer of guardedness.

"I'd like to ask about you. To get some background for the article." She said.

Although a bit uncomfortable with this, he complied. He told her about his years of postdoctoral work in the US and about coming to the UK to set up his own lab. With a flash of inspiration, he started talking about other accomplishments of the Codon institute, selecting carefully, treading carefully. But it didn't work.

"I mean, before that. In Denmark."

He stared at her.

"I did my PhD here in the UK, at Cambridge." He said. "Not in Denmark."

"But you are Danish?"

"Yes, I am," he switched to Danish, to make it obvious. "I am Danish and I grew up there. But my scientific career has be here and in the US. I left Denmark early on." He realized he was being unnecessarily short with her. But this was supposed to be about science, not personal. He wondered how to continue, if at all.

Tina was looking down, at her hands, twisting them.

"You are my father." She said, very softly.

"What?" He said, way too loudly. He stood up, abruptly. He moved behind his chair, held on to the back of it and stood there, staring at her bent head. "Nonsense." He added, with a small shake of his head.

"My mother told me." She looked up, with an expression of half fear, half defiance.

"But that's impossible." He had switched back to English. "I've been abroad my whole adult life. You are, what, twenty? I would have known if I had a kid." He hit the back of his chair. "My God. What a crazy thing to say."

"But she swore it was true." Tina looked at her hands again. "She said that you never knew you had become a father. She never told you." She looked up again, briefly. "And I'm twenty-four."

"It's simply not possible. I am *not* your father." He didn't move. "Who is your mother?"

Tina got up from her chair and collected her things, not looking at him and not answering.

"Just tell me who she is. I'll call her and get it sorted out. She shouldn't be telling you crazy stuff like this." He hit the back of his chair again, but softer this time. "Let's just get this nonsense cleared up."

Tina looked straight at him, blinking fast. He could see tears on her cheeks.

"Look, I'm sorry, but. . ." He started. She was already out the door.

———

He didn't like taking the tube anyway. The confined space, the crowds, the pushing and the not-looking—and now the heat, as well. A long walk home, along Friday-busy streets, was much better. He untucked his shirt and undid an extra button as he settled on a good pace. He scanned people, places and things as he passed. A Goth-style youngster struggling with a rough-looking mannequin in a shop window; the clothes were ridiculous but probably expensive. A Nero and three sandwich places, all close together; how did they manage? The Babylon of languages all around him. The constant onslaught of new faces.

After a while, his thoughts returned to the girl, as he knew they would. It was ridiculous, of course, her claim. He could not possibly be her father. What was she after? Was this why she had wanted to talk to him in the first place? Was there even an Eden newsletter? There was an Eden group, or at least a collection of protestors. And there was this girl, Danish obviously. She had known, or found out, that he was Danish. He wasn't famous, not really. He wasn't rich, so money couldn't be the reason. Had the girl come up with it herself? No, she looked hurt, surprised and hurt, by his outburst. The mother had to be the key, whoever she was. Peter caught glances from a few of the patrons outside a crowded pub. He must look strange, somehow. But to have someone lay claim to you like that, out of the blue. . . Who wouldn't be rattled? He decided to stop at the next pub for a quick one, before the home stretch.

It took him ages to get a beer. It wasn't even what he had asked for, but he appreciated it anyway. It was almost gone now and he was feeling calmer. So, the girl, Tina, had said she was twenty-four. He calculated backwards to get a timeline. The PhD work in Cambridge, the PhD defense, and finally, going to San Diego for his postdoc. He swallowed hard. He had been in Denmark, briefly, in between. He'd worked in his Master's thesis lab, in the Biochemistry department, until the visa and the fellowship came through. A few months, that was all. But he hadn't been seeing anyone, had he? No. Something casual, then? He'd been, what, twenty-seven? Had there been parties, nights out, opportunities to have casual sex? Maybe. Unprotected sex? No. He was too careful for that. But was it possible? Theoretically, perhaps. But he had no memory of sleeping with anyone. No one came to mind. If fact, that whole in-between time seemed a blank. This shocked him a bit, not remembering half a year of his life, even if it was a long time ago. But if this had happened, *if* it had, why hadn't she told him? Tina's mother. And who was she? He had no idea.

He turned into a street of terraced houses. Well-worn houses in faded red brick, the two black stripes of continuity a little surprise, the gap for lower-ground access, the tall white windows. He saw their door. Should he tell Jessie? Yes, of course. It was before her time. But say what? He didn't really know anything. This could be a mistake, a misunderstanding, or a hoax. He would tell her anyway—tell her about his crazy day.

As soon as he opened the door, a familiar laugh floated up from downstairs. He had forgotten. Hans and Alessandra, Nigel and Beatrice; that was tonight. And he was late.

He took three steps down and looked around the corner. They were all still outside, having drinks. He looked at the dining room table. Bowls covered with foil and a large paper bag with a familiar logo. He remembered that they had already done their part, last night, and he relaxed. Jessie spotted him and waved him out.

"Hi, everyone. Sorry for the delay. I, um. . ." He shook Nigel's outstretched hand. Alessandra got a half-hug and Jessie a proper kiss. Still-chuckling Hans got a hand on the upper arm.

"And say hello to this little angel." Beatrice said. She was holding a baby. Alessandra looked like she wanted to take him back, but restrained herself.

"Peter, meet our Leo." Hans said. "Quite the little charmer, but too young for a sitter."

"Well, hello Leo." Peter said, and touched one of the tiny hands. The miniature fingers were so amazingly perfect. He was trying to think of something more appropriate to compliment, when Jessie saved him. She ruffled his hair and told him to go change his shirt.

—

"We are just so worried about her. We don't know what to do."

Beatrice was talking. Alessandra had just returned to the room and sat down with a heavy sigh. She fiddled with something and put it down next to her half-full plate. It was made of pale blue plastic and had rabbit ears: the baby-monitor. They had put Leo upstairs some time ago, but apparently sleeping wasn't his strong suit.

"They all do something, all the girls in her class," Beatrice continued, "self-harm or throwing up or starving themselves."

"You said, what, forty kilos?" Jessie said. "That sounds serious."

"That's what the doctor says. Clinical anorexia. But Catherine won't talk about it. And she won't eat anything I cook—just tuna in water, blanched vegetables, that kind of thing. She says she's healthy." Beatrice sighed. "It's all because of those awful, skinny models."

"Well. . ." Nigel started.

"Have you seen them?" Beatrice said, angrily. "Arms like twigs, collarbones and hipbones sticking out. Ever since we've had these skinny models, the number of girls with anorexia has increased. A lot. That's real numbers."

"A correlation, yes, but not evidence of causality. As a scientist, you know that you can't just. . ."

"Well, I'm not the real scientist here, am I? I'm just the scientist's helper." She turned from Nigel to the rest of the table with a forced smile. "Just ask Beatrice to help you, Nigel says to all his students. Lab Mum. If you only knew. . ."

"Beatrice, please. . ." There was an unmistakable edge in his voice.

"Let her talk." Alessandra said, sharply, to Nigel.

They all sat frozen for a moment, no one saying anything.

"Maybe it's worth getting Catherine to talk to someone who knows about these things." Jessie tried. "Sometimes it's easier . . ."

Peter got up and started clearing the plates. Better to retreat in case deflecting the topic didn't work. He had always thought hiring your spouse as an overqualified lab-manager was asking for trouble, but had no desire to get in the middle of their fight. He moved around the table, wordlessly. Jessie had turned toward Beatrice and was nodding in apparent sympathy with what she was saying. Nigel was listening, but staying quiet. Alessandra was staring at the baby-monitor, looking gloomy. Hans jumped up, followed Peter to the kitchen with two bowls and went back for more. Soon the kitchen area was full. Peter started the hot water running, grabbed a dishtowel from the hook and tossed it at Hans, who caught it midsentence. Hans was talking about the upcoming review. He had been urging Gerald to change the format and was trying to convince Peter to support him. Peter didn't actually think the format was all that important, but he was happy with Hans as company. They could hear the voices from the table, so they kept theirs low. But the chat didn't last for long.

"Hans," Alessandra said, leaning across the counter, "could you go check on Leo?" Her tone was testy and slightly defensive, as if expecting to be disappointed. "You may need to sit with him for a while. He's still awake." Muffled sounds were coming from the baby-monitor in her hand.

"He's probably fine, just babbling a bit. Maybe if we just. . ." Hans started, but Alessandra stopped him with a look.

"I've been sitting with him for almost an hour." She said. "I've fed him. I spent most of the day with him. I'd like a bit of adult company for once. . ."

"Of course. I'll go." Hans turned to Peter with an apologetic look and hung the dishtowel back on the hook. They both looked very tired this evening, Peter noticed.

"Off you go." Peter said. "I don't need help." He turned to Alessandra and added an extra smile, just to be safe. "Please. Go sit. Have some wine. Really, I'm faster on my own." Only after she had left the kitchen did Peter remember that she was still breast-feeding—so no wine. He seemed to know how to say the wrong thing. Hans did, as well. Peter took his time with the dishes.

When he returned to the table, Hans was still absent. He had probably fallen asleep upstairs. The topic still seemed to be Nigel and Beatrice's daughter, Catherine. Beatrice was doing most of the talking. That much he had been able to tell from the kitchen. Nigel looked bored, and when he noticed Peter, sent him a pleading look. Jessie was wearing a patient but strained smile. Alessandra had not succeeded in getting much adult conversation, it seemed, and was now wrapped up in her own thoughts, possibly angry. Maybe he should have come back earlier.

"I hereby declare a moratorium on trying to understanding any behavior exhibited by any teenager—" he said, gaining a half-smile from Nigel "—and an end to any discussion of starvation." He had guessed correctly; Jessie looked relieved as well. "We are about to enjoy our exquisite Glaces Maison. Now, on the left you have glace au chocolat et. . ." He entertained them with this poor French as he created a plate for each of them in turn, adding berries, biscuits and a few nuts with a flourish. He teased Alessandra about ice cream being essential for happiness. She roused herself and joined the conversation. Hans would get melted leftovers, if any.

———

"Oh, God. I thought they would never leave."

"It's not that late, is it?"

"Almost one."

"Really?"

Jessie picked up her small glass of brandy. It was still half full, as was Peter's. He handed his over. She put them both on the usual shelf, by the stairs.

"We are so predictable, aren't we?" She laughed. "Our little habits."

"I think they are lovely habits."

"They are." She said, turning back around. "Very lovely habits for a very happy life." She smiled and kissed him. "But the evening was alright, wasn't it? Considering?" He nodded. They started clearing the table. "I mean, Nigel and Beatrice are never the easiest. But Beatrice was so wrapped up in Catherine this time, we could get her away from the usual complaints."

"Her husband, first and foremost."

"Well, he is a bit. . . overbearing, isn't he?"

"He is."

"I understand her frustration. He's the boss and always the senior author, even though they both have PhDs. Being locked in this subordinate role at work, at forty-something. . . It's got to be tough."

"But it does work. They run a very productive lab. Two experienced scientists working together. I'm sure they wouldn't do as well separately."

"Yes, but. . ."

"You are right. It can't be easy for her."

"Well, I guess they chose it." She sighed. "And each other. He's just so—so stiff. And she's so whiney." She looked at Peter. "How does it go? Saves two other marriages?" He smiled. "I still think that he should be more. . . Well, never mind. Other people's husbands—I don't know how they stand them."

"Whereas I. . ."

"Whereas you are the most wonderful husband imaginable." She put down the plates she was carrying, came around the table, kissed him and wrapped her arms tightly around him. She held the hug for a while. "You know I mean that, don't you?"

"Even when I forget our friends are coming over and I stumble in late?"

"Stumble, that's right—I did smell beer on you. Long day then?" She asked, teasingly. He looked uncertain and didn't answer immediately. "Yes, even then." She added, smiling. "I'm not fooled by a bit of absentmindedness. I know you love me."

"Bunches."

"Bunches."

They continued tidying up, wordlessly coordinated. Peter collected glasses and plates from the patio and closed up to the outside, while Jessie finished the dining area. In the kitchen, the dishwasher was getting serious. Peter took another bottled water from the fridge, Jessie picked up their brandy glasses from the shelf and they went up the stairs. They stopped by the doorway to the guest room. This was where Leo had been sleeping—or not sleeping—for the evening.

"I haven't seen them quite so tense before." Jessie said.

"Alessandra was spoiling for a fight."

"I think she's just tired. Tired-tired and tired of being around a babbling baby all day. I mean, with her intellect and her normal mode of life, can you imagine that?"

"No, I guess I can't. She probably needs to get back to work. Get her old self back. Get her sanity back. Yell at the lazy students instead of her imperfect husband."

"She doesn't yell at them, does she?"

"No, I don't think she does. As far as I know, the students love her. But the barbs were heavy tonight."

"They were." She shook her head and handed Peter her glass so she could adjust the sofa-bed to its normal state.

"She might be worried about losing ground." Peter said. "As they were leaving she asked me if I'd be willing to have a look at a grant application of hers that failed last year. She has to try again soon."

"It hasn't been that long, has it? Her maternity leave?"

"I'm not sure. She's primarily at the university now, not at the institute. But Leo is, what? Four months?"

"Something like that. Maybe Hans can do more to help?'"

"I think he tries. He told me he's always being told off for not doing things the right way. Like with his first wife."

"Oh, Marjorie. Now that was a match not made in heaven. For as long as we have known them, they have fought about those kids. Before and after the divorce."

"And the kids, in turn, have made it hard for Alessandra. The older one, the girl, in particular."

"Well, I suppose Hans knew what he was getting into, having more kids."

"I suppose." Peter shrugged.

She retrieved her glass and they moved on down the hallway.

"They seemed so happy together before—well—before the baby. I hope it's just a phase. I hope they'll make it, you know?"

"I know. I do too."

They put their glasses, and the water, down on the low table and collapsed with parallel sighs into the large, very old and very soft sofa. They always ended up here, for winding down coziness.

"God, we are so lucky." She exclaimed dreamily. He was in his usual corner spot. She snuggled closer and he wrapped an arm around her. "I am so lucky. I got the best one. I love you so much, you know that?"

"I know that. I love you too." He smiled above her head.

"Alright, then." She said, softly and tiredly. It had been a very long day: The endless crap at the journal, everyone wanting her attention before her trip and everything feeling like damage control. Normally she'd want to talk about it and get it out of her system. But not tonight. Tonight it was not needed. "We are so lucky," she repeated.

Her eyes closed ever so slowly. Her breathing became deeper, lifting his arm gently, regularly. He smiled contentedly to himself. She was right, he thought. They were lucky, very lucky. He tried not to move—for as long as he could manage.

Chapter 3

Monday morning, the weather was still on high summer and they were still there: the animal outfits, the cages, the placards and the chanting whenever the streets got busy.

The following couple of days, Peter avoided them by coming in early and staying late. He had plenty to do. Top of the list was reading and scoring fifty-three grant applications for the BNC and finalizing Philip's paper. It was a good story but Philip never got around to finishing it. Peter had asked Susan, his long-term lab technician, to help with the final experiments. Focusing on the paper, however, proved difficult. He was distracted. He kept expecting a call from the front desk, saying he had a visitor. It didn't come. After a while, he realized that he had scared her off. If he wanted to know more, he would have to approach her.

Thursday, he ventured out for lunch and looked closely as he returned. She was still there, but she was not looking in his direction. Maybe that was a sign, he thought. Her outrageous claim would just fade away. He reckoned it was a good thing that he never got around to telling Jessie. She didn't need more to worry about right now. Saturday had been way too busy for a proper discussion. She had had a million things to wrap up before she left. Sunday morning was even worse. She had been so excited and even a bit nervous. And he understood. They didn't talk about Oak Hill, either. It was sensible. Better to wait until she knew more. And he knew more. Then they could talk properly about the future and their options. He almost convinced himself.

———

Peter was going up, Gerald was going down.

"Peter—" Gerald said. It was not just a greeting; it was an opening. They both moved to the wood-clad side of the staircase. The other side, open to the central space and to the floors above, did not invite lingering. Everyone agreed that the wide central staircase with its attractive teakwood steps was a major design success of this building. It was good for everyone's health and for unscheduled, brief conversations.

"So—how did the interview with the animal rights people go?" He nodded toward the entrance. From where they stood, they could make out their backs but not the cages. "Last Friday, wasn't it?"

"OK, I think. I haven't heard from them since."

"But did you have a good feeling about it? Were they listening?"

"I'm not sure I could tell, really. And it was just the one girl."

"Did she seem hostile?"

"More interested than hostile."

"It would be excellent if we could make them move on, somehow. There are far more appropriate targets for their—their displeasure. Not that I wish problems on anyone else, but. . . Bill has been bending my ear, insisting I do something. Hoping for trouble, no doubt." Gerald glanced around quickly. No one was close. "Now Carol has started as well." He sighed.

"Carol? Really?"

"Apparently her postdoc, the Chinese girl. . ."

"Huifen"

"It seems she feels intimidated by them. She complained to Carol. Carol talked to me and. . . Ultimately, I'm responsible for everyone's wellbeing, their safety and so on." He trailed off. "So, I was wondering—since you've met one of them—could you find out if there is anything we can do—short of closing the institute," he scoffed, "that could end this? Peacefully, obviously."

"I can ask. But I don't know whether. . ."

"I have an idea." Gerald smiled confidently. "How about you meet this girl again, your contact, but this time with Darya? She is closer to their age, female, an extrovert of sorts and always enthusiastic about her work."

"Darya is a wonderful scientist, but. . ."

"And she works on C. elegans. No one is concerned about the life and welfare of a few worms. Maybe she can connect with them?" He repeated the smile of before. "What do you think? Give it a try? Ask Darya."

"OK, sure."

So that was settled, then. He would try to talk to her again. Having Darya there might help with the initial awkwardness. But he would need to talk to her alone as well. He needed to know more. While contemplating this, he proceeded slowly up the steps. On the second floor, he saw Darya in the common area with one of her PhD students. Peter recognized the student, a shy, bright one, but he couldn't recall the name. They were both looking intently at a laptop, the screen of which was facing Peter. He saw graphs in the background and what looked like a movie in the foreground. He stepped closer so he could see better and then waited until the short movie had looped twice. From the flashes, he guessed local photo-activation. The response, shown in false-color intensity and going up and down in a complex pattern, was a sensor of some kind.

"This is amazing, absolutely amazing. Well done, you." Darya beamed at the student, whose blush was only partially disguised. He looked up at Peter, which made Darya turn around.

"Peter, you have *got* to see this." She said. "Nihal has just found something absolutely fantastic. Nihal, you tell him."

Nihal looked down, hesitating.

"OK, OK, I'll tell him." She turned to face Peter. "You remember the chemotaxis circuit we have been studying. . ."

"Sure. . ." He did. He enjoyed her frequent visits to his office. She rarely came by to complain or to ask for help with grants and papers, as most of the others did. She came by because she just *had* to tell him about their latest result. It was too exciting to keep to herself. She had that look now. Running the movie once more, she started to tell him. Quite quickly, he realized that he was not paying proper attention.

"Darya, I'd love to hear the rest of it, I really would, but a bit later. I need to ask you for a favor first." He paused and Darya closed the laptop. This gave Nihal his exit signal. He took the laptop from the desk and glanced at Darya. She gave him an encouraging smile and a brief nod and he moved off.

Peter explained what Gerald had proposed. She shrugged and agreed without further comment. She was around, she said, so any time would be fine.

"Thanks." Peter said. "I'll let you know what time—if she agrees."

"No problem." Darya stood, picking up her cup and a few loose papers.

"I'll catch you later on that result of Nihal's, OK?" Peter added. "It sounds really interesting."

She smiled broadly, adding a quick shudder of excitement and walked off. He resisted going around to his office. Instead, he went back down the stairs.

As he approached the group, someone moved toward him. Peter recognized the man called Alistair from the first meeting. He did not see Tina now, although she might have been behind a mask.

"I was wondering if I might speak to the girl, the woman, who interviewed me last week. Tina, I think. Is she. . ."

"She's not here." Alistair said, his voice hard, unpleasant.

"But. . ."

"She left for the day." An ambiguous half-smile appeared on Alistair's face and his voice caught a bit of sing-song—mocking, it seemed. "Are you interested in our activities, perhaps? Something I can help you with?"

"We didn't. . . the interview wasn't quite finished. And there's another scientist who would like to talk to her, to your group. She. . . She uses worms to study behavior. She's great." Alistair waited for more. "We would like to explain more about what we do here so that Tina—or you—can pass it on." He felt like he was begging. This was totally wrong. He should just walk away.

"We appreciate that you take an interest in Eden and in our mission." Alistair said with the mocking half-smile of before. "I'll let Tina know."

Peter couldn't tell if the sarcasm meant that he would do no such thing. But he had done what he could for now. He nodded goodbye and went back inside.

———

She showed up the next day. It was the same routine as last time: no I.D., no last name and stern looks from the security guard. But something had changed. This time, Peter was studying her much more closely. He tried to be discreet about it, one glance at a time. He noticed her rounded, almost lobeless ears, then her eyes, greyish with flecks of warmer brown and finally the movement of her hand as she secured a lock of loose hair behind her left ear. Darya came down the stairs just then and he introduced them. He watched as they exchanged a few remarks. Darya was all smiles and already going a mile a minute, as he had hoped she would.

The three of them went upstairs to the common area and Darya pulled out her laptop. Tina pulled out hers. Tina started with questions very similar to those she had asked Peter. She immediately got some very detailed answers from Darya. Too detailed, too technical, Peter thought and he jumped in with extra explanations. But the movies were a success. Tina started asking about what she was looking at, clearly taken with it. Peter added questions as well. The conversation was pleasant and non-confrontational. No one mentioned furry animals. It was working. Maybe. Finally Darya closed her laptop and got up from her seat.

"Look, I have to go. There's a seminar on soon and I can't miss it." She looked intently at Peter.

He had forgotten. "Yes, we... I'm sorry but. . ." he mumbled as he got up. Tina followed suit. They were all standing.

"It was a pleasure to meet you." Darya said, pleasantly, and stuck out her hand. Tina shook it with a shy smile. "I hope you can convince your friends out there that we are not the enemy. We really aren't." Darya smiled, turned around and headed off with her usual briskness. Peter stayed behind. He and Tina both looked at the disappearing Darya, not at each other. A few moments passed.

"Look, we should talk." Peter finally said. "About... About what you said... About what your mother. . ." He took a deep breath. "I would like to talk. If you're willing."

"OK." She said, looking down.

"And if. . . if there is any way you could convince your group to give it a rest, that would be super helpful. Really." She looked up and directly into his eyes. He felt a small shock. He wasn't sure what from, or even whether it was pleasant or unpleasant.

"OK." She repeated.

They were both quiet again.

"So maybe Monday?" He said. "Or Tuesday? I'll be here."

She smiled, tentatively, but genuinely, he thought. With another "OK" she straightened up, gave the ponytail a good swing and hurried down the steps. He moved closer the stairs and looked down. She was wearing a loose cotton skirt again and it billowed around her as she descended. She did not look up.

———

Overnight, the weather broke and normal English summer returned, cool and overcast, with occasional rain. Peter welcomed it, but it meant he could not sit outside as he had planned. The wicker chair was wet, Jessie's swing as well. He sat on a barstool by the kitchen instead. He soon realized the Saturday morning ritual of slow coffee and croissants from around the corner wasn't doing any good. It was too quiet without Jessie. He picked up his phone and sent a text to Hans. "Jog and pint this pm?" Then he remembered the baby and added another text: "If no time, no problem." He scrolled through the contact list, pausing at a few, but finally gave up. Better another quiet night with take-out, a couple of beers, Netflix going and Skype on and waiting. Jessie still hadn't met with Tony and he still hadn't really talked to Tina, so they would just have their usual, meandering chat, filled with details from the day. He smiled as he imagined it.

He decided to put in some hours at the lab, possibly followed by a long walk. It was a good distraction, city walking, casually looking at people and things. He'd get a cup of coffee somewhere. Hans had begged off, as expected. He finished his breakfast while glancing through the news online. Nothing engaging. Then he got up briskly and went over to the glass door to lock up before leaving. He looked up. As he had hoped, the sky was presently unthreatening. Standing there, he remembered something else: That outside lamp they had seen a few weeks ago. Jessie had loved it. He wasn't sure which store it was, but he remembered the area they had been walking in that day. And he had all weekend to find it. It would be a surprise. Thus cheered, he picked up his phone, his jacket, an umbrella, just in case, and headed for the door.

—

Approaching the institute on Monday morning, Peter let out a sigh of relief. No protestors. They hadn't been there over the weekend either, but he hadn't been sure whether that was part of the normal pattern or a real change. It seemed the latter.

"Looks like we succeeded." Darya said when they met at the coffee machine.

"Or it's just the change in the weather." Peter said. "Fair-weather protestors?"

"Nah—" she shook her head "we succeeded. Gerald has already thanked me. He said to pass it on. He is terribly grateful."

"Well, good." He smiled. Her slight mockery of Gerald, the fake posh accent, was good-natured. "So what about those new results of Nihal's? I'm sorry I was too distracted on Friday. Whenever you have time. . ."

"Now?" She held up her laptop. "Always prepared."

"Now is good."

They found a free table and got started. Peter already knew the background. Darya had noticed some time ago that one particular cell in the chemotaxis circuit she was studying had robust and persistent calcium waves, but their frequency varied from animal to animal. She had shown him plenty of movies to convince him of that. The variation could have been random, or irrelevant, but Darya had been sure there was more to it. She had encouraged Nihal to investigate it. He had been systematic and tested lots of variables. It had paid off. They had found out that the frequency of the waves correlated with specific aspects of the animal's recent experiences and, intriguingly, could be modified by training. It seemed to be a form of memory. Darya took Peter through the numbers. They were solid. Clearly, Nihal was a diligent student. They were lucky with the quality of the PhD students here, Peter reminded himself. He thought of Mihai. Julia showed the same promise. Philip—well— sometimes, one couldn't tell.

"I'm impressed." He said, when she paused. She flashed a quick smile at that. He *was* impressed, perhaps even a tiny bit envious. From when they first interviewed Darya, he had been one of her strongest supporters. Her work was so interesting, so innovative. That she worked on an invertebrate model system was, admittedly, another plus. They understood one another. He thought of all the effort she had put into setting up this system. Now it was paying off. She deserved it.

"Ah," she said, with added drama, "but I still haven't told you the best part. This is so super-cool. It comes from our collaboration with Hans." She explained further. Peter listened. After a while, she slowed down again and smiled. "It all started with a casual chat over coffee. In this exact spot, actually." She pointed at the desk before them. "We were talking about how information is stored in biological systems versus in computers, short-term and long-term." She sighed, happily. "It's this place, it's just so. . ." with another big smile, Darya swept her arm across the central space. In so doing, she seemed to have noticed the large wall clock.

"Sorry, sorry, I have to run." She got up. "An eager undergraduate will be at my office in a few minutes. I'd better be there."

Before he had time to respond, she was off down the corridor, her brisk steps doing double-duty. Peter got up more leisurely and smiled to himself. Had he had that much energy at her age—ten, fifteen, years ago? He hoped so. He thought so. And he was still doing fine, he reminded himself.

———

This time, she was wearing blue jeans and bright green raincoat. She had not brought her laptop. The sign-in issues were the same as before, though, and Peter had to reassure the guard again. They didn't speak directly until they had reached his office and closed the door. It wasn't a hostile silence, Peter thought, more an anticipatory one.

"So," he finally said, "you really won't tell me who your mother is?"

Tina shook her head, no.

"And not your last name, either?"

She shook her head again, but this time with the hint of a smile. Did she want to start a guessing game, he wondered, or was she just acknowledging his not-so-subtle indirect approach? The latter, most likely.

"You understand why that makes it difficult for me, don't you?" She nodded. He continued. "This is a huge thing. And all I have is your word, your belief."

She nodded again. She seemed calm, very far from the upset of the first time they talked about it. She must have noticed the change in his attitude. He had noticed it himself. Where did it come from? That movement she had, flicking her hair behind her left ear? Maybe. But he wasn't giving in so easily.

"OK," he said, "so how about we trade information? You are twenty-four?" She nodded. "After we first spoke about this, I realized that I was, in fact, in Denmark for a few months, twenty-five years ago. So I suppose what you say is possible. In theory."

"You see?" Her smile was so spontaneous and so complete that he almost lost all resistance right then. But he needed more.

"It's just—I don't think I was with anyone. I don't make a habit of sleeping around." He paused. She shrugged. "So without knowing who your mother is, I'm sort of stuck." He paused again. "She told you who I was. Doesn't it feel fair to reciprocate?"

"I guess." She said, followed by a longish pause. "But she wasn't telling me to be fair. She has been keeping it from me, lying, for my whole life. I made her tell me."

"Made her? How?" She didn't answer. "Look, maybe it wasn't fair of her to keep it from you. But this is just as unfair. Aren't you doing to me what she did to you?" She wavered, he could tell, but ultimately held her ground.

"It's not the same. You don't know what she's like. I can't..." Tears were collecting.

"OK, OK, I'll stop." He almost reached out a hand to her, but held back. She recovered. He spoke again, calmly. "So, you are from Copenhagen?"

"Yes." She admitted.

"And is this where you got started with the animal rights stuff?" She eyed him carefully. "Don't worry, I'm not trying to get information about all that." He continued. "The institute is very grateful that you, the Eden group, decided to move on. Thanks for your help with that."

"You are welcome." She said, surprisingly softly.

"Anyway, it's just me and you here—that's all." She didn't add anything. He continued. "And the flies, of course. Flies are, well, you know, flies. Irritating buggers." He smiled gently, probing. She reciprocated. "Did you know that the eye color mutants of fruit flies represent some of the first genes to be studied properly? This was back when genes were just a theoretical. . ." She looked puzzled. "Well, I suppose I've already told you all that." He added, still unsure how to proceed. She saved him.

"The woman with the fluorescent worms—she was awesome. Seeing all that stuff happening. . . It was fantastic." So Darya had indeed made a big impression. Gerald was smart about such things.

"I agree—Darya—Darya's work *is* awesome. We do our best to learn as much as we can from these simple critters." He smiled. "But, scientists do have to look at animals that are more like us sometimes. Or we won't have any new drugs." He took her through a couple of good examples. Testing penicillin in mice—and cisplatinum, to cure cancers. He had prepared himself well.

"I understand that. But that doesn't excuse people torturing rabbits to make new make-up."

"No. I'd tend to agree with you on that."

"There's also no excuse for mega-farms that keep animals under terrible conditions just to produce more and more meat, milk and eggs, or to make it so incredibly cheap that fat Europeans and fat Americans can eat lots of it, every day. It's disgusting." She looked angry. She meant it. "Millions of animals suffer and suffer—for cheaper food. Have you ever been to a high-intensity dairy farm?"

"No, I guess I haven't, but. . ."

"I have. That's why I'm a vegan. Do you know that they take the calves from their mothers as soon as they are born? Then they milk the mothers to an early death in tiny cubicles. And the calves—sometimes they kill them straight away, sometimes. . ." She continued explaining, arguing her case. He didn't stop her or contradict her. For the most part, he found that he didn't actually disagree. In truth, he hadn't really thought much about it before. He also enjoyed listening to her. In addition to the anger, there was passion and, as he had sensed previously, intelligence. When she stopped talking, she looked down. Maybe she hadn't intended to say so much.

"I understand what you are saying and I can see that you really care. But why aren't you protesting at these farms, then, instead of at a research institute?"

"I have done, I. . ." She stopped, sensibly. "Well, the Eden group is—" She stopped again. Finally, she settled on a direction. "That's just as wrong. Systematic torture of animals just to. . ."

He let her talk for a while longer, half clichés and half the history of animal rights actions. She seemed a bit less passionate now, and less certain.

"Did you visit farms in Denmark? Is that how you got interested in this? Do your grandparents live in the country, perhaps?"

He used the Danish words for maternal grandparents, hoping to strike a connection. Much to his surprise, it worked. A giant smile lit up her face.

"Mormor and Morfar had these rabbits." She started eagerly, switching into Danish. She told him about the rabbits and in particular about one called Pelle, her favorite. Inside the large rabbit-pen, Morfar had built a tiny house, where she used to hide when she was visiting and didn't want to go home. There was no mention of other animals and the visits seemed a frequent occurrence, so he guessed her grandparents must have lived close by. He asked a few questions, also in Danish, but she stopped her tale soon after, switching back to English.

"Alistair says keeping pets like that, in captivity, is also a form of torture. It's sentimental and not natural. . ." She trailed off. "I know that he's right. But I was just a kid. . ."

He nodded, thinking of how to steer her back to talking about her grandparents, without appearing to be trying to hard. Nothing came to mind. Some more obvious questions did, however.

"Do you watch nature shows?" He said, a bit brusquely. "Documentaries about life out there, in the real, natural world?" She shrugged, probably guessing where this was going. "Do you know how animals die out there?" He knew he shouldn't be saying this, not now, but he couldn't help himself. Perhaps it was the mention of Alistair that had provoked him.

"But at least they are free before then. Not imprisoned. And they have their babies naturally. Alistair says. . ."

"Have you watched a cat slowly killing a mouse? Lab-life is not bad, by comparison. Whatever animals we have here, at least they are treated humanely and killed humanely." Talking so directly about killing was obviously a mistake. She withdrew fully. He tried backtracking, asking about her life in Copenhagen and in London. But it was too late. Her answers were brief and uninformative. The former liveliness seemed impossible to rekindle. The switch frustrated him. Not knowing how to undo it, how to get her to respond, frustrated him even further.

"We have to finish up now, I'm afraid." He finally said, acknowledging defeat. But there was still so much he didn't know, he realized. He had less than a week before Jessie got back. "But I would like for us to talk again." He added. "If you want to." She did, apparently. They arranged for her to come back two days later.

On their way out, he remembered how much Tina had liked Darya's movies. On a whim, he decided to show her the fly room. He had mentioned the eye color mutants. Maybe a little show and tell could help erase the misplaced words.

To get to the fly room, they had to go through the wet lab. It was lunchtime and the space was empty, as he had hoped it would be. They walked past benches filled with racks, tubes, pipettes and notes. Each space was personalized by use and habit, by labels and Post-Its. Tina was looking around, seemingly curious. Passing through two more doors, they entered his small fly room. He would have liked a bigger, shared fly room, as he had had in the old building. Students from different labs talked while working, occasionally even about science. It was nice. But here it would have been impractical. Lab placement at the Codon was organized around scientific questions rather than choice of model organism. Perhaps it was for the best, he thought, and now he was used to it. The room had four fly stations, one of which was occupied.

Mihai, of course.

"What's that smell?" Tina exclaimed in surprise.

"It's the fly food, basically. Some people can't stand it."

"No, it's not bad. It's just—strange." She volunteered a weak smile.

"After a while, it's the smell of home." Mihai said, turning around and grinning, full of charm, at the newcomer. He probably thought she was a new student, looking for a placement.

"Mihai, Tina. Tina, Mihai." Peter introduced, while indicating with his hand. Minimal information seemed the best option at this point. "Maybe Mihai can show you some flies—and a few of those eye color mutants I mentioned." He looked at Tina, questioning. "If you'd like." She nodded. "And if you have time." He looked at Mihai. He nodded as well. "If you're lucky, Mihai might even show you this fantastic little device he has developed." Peter continued.

"Me and Victor." Mihai corrected him.

"Yes, of course. Mihai and his fellow PhD student have developed this neat device that tracks individual flies. With it they can identify any fly that remembers its training better than others. Or worse." Tina looked unsure. "What we talked about last time? Our work on memory in flies?"

"Yes, I remember." She looked like she did. Peter continued. "Mihai has found a very interesting mutant. He just hasn't told me what it is yet." Mihai looked proud and a bit mischievous. "I'm getting ever more curious." Peter added, looking intently at Mihai. "When will. . ."

"Still three months to go. I'll tell you as soon as I'm sure."

"Hmmm." Peter shook his head. "Well, maybe I'll just leave you to it."

"Sure." Tina said, smiling now, but more at Mihai than at Peter.

"No problem." Mihai added, returning a slightly cheeky smile.

Tina was quite attractive, Peter realized. But Mihai was a good person, a decent person, Peter knew, so he did not need to worry. Mihai anesthetized some flies and put them on the pad. He started explaining some simple features, the eyes, the wings and the body shape, and offered Tina the eyepieces.

"It's all a blur." She said, squinting. He adjusted the eyepieces to her eyes and rechecked the focus for her. "Oh, my God," she exclaimed when she looked back down "they're huge. Are those really eyes?"

Peter realized he was superfluous for the moment. "I'll be right in here." He said, nodding toward his office. There was no response. They were already deep into appreciation of Drosophila eye color mutants.

From his office, Peter had a good view of the fly room via a large, internal window. He kept a careful, but discreet, eye on Mihai's demonstration. He wondered whether Tina would tell him who she was. Mihai might have recognized her from outside, the week before. No harm in that, he figured. Surely she wouldn't say everything.

A little while later, Susan came back from lunch. She took her usual place in the fly room and it seemed she greeted Mihai and the visitor without further comments or questions. He was surprised not to see Julia yet, but then remembered that she had a course at the University.

Finally, Ilana came back. She sat down at her desk and woke up her computer. Peter knew there was a straight line of sight into the fly room from her desk. A few moments later, Ilana looked up, then got up quickly and marched purposefully toward the fly room. Peter almost jumped from his chair and reached the door at the same time as her.

"Ilana, Tina. Tina, Ilana." He introduced them with clear, deliberate gestures. They both smiled, but without warmth. "Tina is with the animal rights group that has been taking an interest in our institute." He said. "She wanted to understand how we study brain function and memory using flies, without harming higher animals. She plans to write about it for their newsletter. I told her about our work and Mihai was kind enough to show her some flies."

Peter looked straight at Ilana, making sure they had eye contact, then did the same with Tina. Both seemed to be processing what he was saying—and not saying. Neither of them responded directly. Only Mihai looked genuinely surprised. He must not have recognized Tina from the protests. Ilana held out her hand to Tina.

"Nice to meet you." She said, in an artificial monotone. Then she turned to Peter and added. "I have some new data you might want to see." She left the room quickly, apparently not registering Tina's polite response. Peter followed her to her desk, closing the two doors behind them.

"I assume you didn't tell her?" Peter drew a deep breath, but didn't answer immediately. He looked uncomfortable. She continued. "About me, I mean, and my project. I thought my project was important."

"It is. Very. Of course it is."

"But you said you only work on flies."

"Well, I didn't actually... Anyway, what would be the point? I'm trying to..."

"Honesty is the point."

"It's not that..."

"Important?"

"Come on, Ilana. You're an adult. Be reasonable."

Chapter 4

Tony got up from behind his large, neat desk and came around to give her a loose hug of welcome. Jessie appreciated it. When she saw him last year, it had been a slightly awkward handshake. That had been her doing and she had regretted it almost immediately, given how well they used to know one another. It was the shock of how much he had changed that threw her off. He used to carry a few extra pounds. None now. Now he was lean and fit, but oddly desiccated, somehow older than his years. His hair was completely gray, what was left of it. His eyes still had that roaming, piercing alertness that would miss no challenges. But they also looked tired, and, maybe for that reason, kinder.

"Not bad, is it?" He gestured toward the large picture window behind him. It was like a painting: blue sky, sun on a green field and a patch of mature trees, the slope of gentle hills cradling it all.

"Not bad at all." She agreed.

"I'm sorry I have you flying back and forth from coast to coast. But there were a few people I wanted you to meet and they weren't here last week." Tony picked up a single sheet of paper from the corner of his desk and handed it to her.

She glanced at it. "My schedule?"

"Yes. The two of us have the rest of the day today. First, I will give you a tour of our humble premises. . ." He smiled. She smiled. "Then we will talk. Tomorrow," he pointed at the schedule, "you meet Gustav Krüger and Junko Ito. Both are senior hires and in the process of setting up their labs here. I didn't give you their names over the phone. It'll be official next week."

"Gustav I know. Amazing technology. Junko Ito?"

"She's fantastic." He said with emphasis. "Incredibly smart and nice as well. I'm sure you will enjoy talking to her. She models cellular interactions based on membrane 'omics combined with three-D reconstructions of cell shapes and inter-actions. She has only recently switched to focusing on neurons, which is probably why you don't know her."

Jessie nodded and scanned the rest of the short schedule. "Benjamin Harrison?"

"He's head of admin. You're seeing him for half an hour. Just to say hello, essentially." His eyes sparkled. "Now, there's one more senior hire. Mengyao Wang."

"He's coming here? Leaving Stanford?"

"Yes. He'll be joining us next month."

"Obviously, I know Mengyao and his work. Who doesn't? Wow." She shook her head slowly. "But I suspect he might not be too keen on me. We've had quite a few disagreements over the years."

"Mengyao can be a bit of a bully, I suppose. But he respects you tremendously. He told me so, quite unambiguously. He's sorry he can't be here this week."

"So he knows about my visit?"

"Yes, and he knows what my intention is. I've talked it over with Gustav, Junko and Mengyao—my three musketeers. They're all with me on this. Obviously, they are also helping me with the junior hires. We've just selected twenty excellent candidates. They're all amazing. You should see. . ." He grinned and shook his head. "The interviews kick off September first."

"That soon?"

"Yes, things are moving fast now. I hope to have you on board by then, so—"

"Now wait a minute." She said, and frowned. "You are getting way ahead of me here. I still have another job. And I—we live in London."

"Of course, of course." He smiled. "All in good time. First, the grand tour of The Oak Hill Research Center." In a mock-formal way, he gestured toward the door. "I'll tell you about the labs and facilities as we go along. Your mission—should you choose to accept it—" he said, with another flourish of a gesture, "—we will discuss when we return."

She shook her head with an indulgent smile and followed him out the door. The main building was as sleek and elegant on the inside as it was on the outside. Openness and transparency had been the trend in science building architecture for some time now. Apart from the view, the tour reminded Jessie of visiting the Codon Institute when it had first opened. Every room sparkled brightly, all potential. Looking out onto the glories of nature rather than a busy cityscape gave everything a different feel, though. The swaying green was serene, unhurried. One could contemplate the view for hours, she imagined, letting inspiration come rather than be chased. She thought of her tiny office at the journal. That building was well past its prime, a so-called character building with single pane sash windows and scuzzy carpets. She forced herself not to dwell on the comparison. Tony moved through the clutter-free rooms, explaining their plans for sharing of space and encouraging shared projects. They looked at the generous core facilities that were to be kept flexible. Jessie asked lots of questions, mostly short and direct. Tony answered with matching directness. Here and there, the first few people were working, adding color to an office or a lab. Everyone smiled and said Hello.

They continued the tour to the annex, a low, box-like structure with a solid glass front and panoramic windows to the back. It was connected to the main building via an enclosed glass corridor and they entered this way. Tony showed Jessie the meeting rooms first, each a different size but none very big. For two of them, large windows covered the far wall, giving an open feel and visual interest but limited direct sun. She quizzed him about the A/V setups and other technical aspects. Next, they walked through a central space with clusters of sofas at one end and a sparsely decorated, lobby-like area at the other. He gestured to a corridor coming off this space, opposite the meeting rooms.

"Twenty guest rooms up that way. You'll be staying in one of them." She nodded. They went out through the lobby, through the sliding glass doors and stood outside while Tony finished describing the layout of the place. Jessie enjoyed the sunshine for a moment. The inside has been excessively air-conditioned, she thought. She smiled to herself at her acquired sensibilities.

"Shall we go back to my office?"

"Sure."

"So, as you see," he started as he was getting seated, "we have a fantastic site. We have stable funding and a few top notch and very interactive scientists on board. Most importantly, we believe that the best up-and-coming PIs in the field—innovative, imaginative, energetic and ambitious—will chose to join us at Oak Hill."

"Everyone is talking about this place, Tony. Who wouldn't want a job here? Especially the young scientists, starting out on their independent careers."

"Exactly. And because we have all this, I think we also have a duty to go above and beyond, to improve science as much as we can."

"Admirable."

"And one area that needs improvement is publishing."

Jessie was about to interrupt, but Tony held up a hand to stop her.

"Hear me out. When we planned this institute, we asked a lot of eminent scientists for advice and input. But we also asked younger PIs about what they felt limited their ability to get the very best science done. There was funding, of course."

"Of course. Always."

"But the other recurring theme was publishing. The PIs, even the successful ones, felt that the process of getting papers published had become so arduous that it almost killed their desire for doing science."

"I've heard that as well."

"Both groups mentioned recent positive trends: resource papers, archival journals, self-archiving. But for the top stuff, things have not improved. Despite good intentions, the new journals are not that different. Reviewers tend to stick to their habits, editors too."

She nodded. No comment was needed.

"That's why I was so intrigued by the idea you pitched to me last year." He said, leaning back in his chair. "Tell me more about it."

"OK. Right." She took a deep breath. "So—I was at the Franklin Institute, talking to a group of bright and wonderfully irreverent PhD students."

"They suggested it?"

"No, they just asked the right questions. I was giving one of my usual talks about the publication process, the role of reviewers and editors and what we look for in a paper."

"People really appreciate those talks, I've heard."

"I'm happy to hear that." She smiled, briefly. "Anyway, this time I had just had a very stimulating discussion with a young PI. She had made a really tantalizing, unexpected observation and was so excited about this and what it might mean that our discussion went all over the place. She had not decided on the final interpretation yet and did not try to sell it as a finished story. Most people I talk to do."

"Talking to you is an opportunity for them. Prepublication enquiry, in person."

"I get that. And I do listen. But this discussion I had with the young PI, exploring possibilities before her observation had been put into a narrative, it was exhilarating. It really got me thinking."

"That's what good colleagues do, if you are lucky enough to have them. Toss new data and ideas around, see where it leads."

"Sure. But I wondered if it couldn't also become part of the publication process, or, more precisely, one type of publication."

"You said the PhD students inspired you."

"Well, they simply asked why. I explained how a paper was like a story, a true story, of course, but a story, a narrative, nonetheless." Tony was nodding along, half-smiling. "Someone just put up their hand and asked why."

"Why?"

"Why do we use narrative? Of course I answered that narratives are useful because they make sense of facts and put them in an order."

"It's human nature. You. . ."

"Exactly." Jessie said, cutting him short in her eagerness. "But afterwards, I started thinking more seriously about our fixation on narratives—and the consequences. Narratives represent a path. At many points along that path, the author will have made a choice: What to pursue and what not, or how to interpret something. Each choice has to be argued, of course, and backed up by facts and experimental data. But it is still a choice and every choice limits where you can go from there."

"Some more than others, but yes, I see your point."

"You can get locked into a story, once it starts to take shape." He nodded. She continued. "There's also the issue of length. To get into a good journal, stories often end up being very elaborate, with a lot of steps: more details, other organisms, new drugs, whatever. Reviewers notice steps that are less solid and ask for more experiments to strengthen these. Authors comply and the new data have to be interpreted. Based on this, reviewers have more questions. It can go on and on."

"A good paper often takes years to publish these days."

"If we, as editors, try to limit the reviewing process, we catch flak from reviewers. So we let it go on. The authors get frustrated, which is understandable. With so much time and effort going into it, they may also get overly vested in their narrative. This carries its own problems, the hard push to get that last result for the reviewers." Tony nodded, with a resigned expression. Like everyone else, he had been there. "Right," Jessie continued, "but what makes a paper worth top-level attention is often just one observation, one key experimental result. Not always, but surprisingly often. By the time we see this observation, it has been embedded in a story: A leads to B and so on to the conclusion. It occurred to me that science might benefit if we—at least sometimes—got these eye-opening observations out there for the community think about without a constraining narrative."

"OK…"

"So my idea was: How about making a journal in which we publish single, intriguing observations—repeatable, of course, top quality, but just one figure—with an explanation by the author, and—here's the important part—directly attached to it, critical questions, concerns, caveats, interesting implications and references to related findings, all provided by knowledgeable experts."

"And you thought this could be accomplished by…"

"…by having the observation presented at a small, intense meeting." She continued rapidly. "Each observation must be discussed by the participants, experts, in dialog with the author. I think doing it in person is essential. Having an audience will keep the discussion constructive, and people like to show off, so they'll try hard to come up with good ideas. I also think it is important to have the meetings at a place associated with scientific excellence," she smiled, "like Oak Hill." He smiled. "I imagine twelve or fifteen scientists at each meeting, all independent PIs. Each participant comes with one new observation to present—a brief presentation, followed by a longer question and discussion session, where everyone participates. The idea is to publish all comments alongside…

"Attributed to the person commenting by name."

"Exactly. Thoughtful comments will be appreciated. If caveats or alternative explanations are presented, the reader can consider them. Constructive input gets rewarded and people will be too embarrassed to roll out their competitive nastiness.

The comments immediately become part of the public record, but they do not prevent publication. This is key. The observations would be published soon after the meeting and each one would be a citable publication." She paused, caught her breath. "Now, for this to have real impact, the new journal would have to be considered a very desirable place to publish. This is where the institutional backing comes in. If a top funding agency and top scientists in the field support it..." Tony smiled. She hurried on. "What was missing in my initial pitch of the idea was a realistic plan for how to get top scientists to participate from the start. If they do, and they do a good job, the rest will flow naturally. I am sure of it. So if you, the foundation and Oak Hill support it..."

"Then it is suddenly possible."

"Yes. Think about it from an author's point of view: If you have a fantastic observation, you can have a top-level, citable publication in a few weeks or months, not years. Of course, this means the meetings should be reasonably frequent, maybe once a month, shifting subfields, giving a full rotation once a year."

"I like the idea, very much. As you know." Tony said. "But it is a complete departure from the usual paper structure and peer-review model. No wonder your journal wasn't so keen on it."

"It's still peer review, because of the comments, but yes, dramatically different. Review as an addition, not as a gating mechanism. A small revolution." She grinned. "Wouldn't everyone want to be part of that?"

"Well, maybe not everyone." He smiled. "But I would. We would, here at Oak Hill. That's why you are here."

They were quiet for a moment.

"There are bound to be skeptics." Tony continued. "And the approach won't fit all stories—" he smiled and corrected himself "—all scientific discoveries, all papers."

"No, of course not. In some cases, stand-alone observations are not sufficiently informative." She nodded. "Those can go the standard publishing route."

"We need to think hard about how we do this so we capture the best stuff. Practical details may make all the difference."

"Yes. Each PI should go to max one such meeting a year. That way, they will choose their most interesting new observation to present. It should also encourage active commenting. Less meeting fatigue."

"Right—and if the observation presented is not so interesting, that person won't be invited back."

"No. Constructive participation in the discussions can be another criterion." She paused. "Obviously, the critical aspect will be who is invited to present an observation—and who decides who to invite. Because that will be our gating." She fixed Tony with a firm look. "I'd need advice from you and your heavy hitters. I'd need a knowledgeable and very responsive advisory board."

"Of course. You will get all the input you need. When you are Scientific Co-Director of Oak Hill, people will be happy to help you." He paused to let the new job-title sink in, then continued. "You would of course be responsible for this new venture. You would also be involved in many other decisions around here, including hiring of junior group leaders. It will be a very interesting job."

"Very. And I'm very flattered."

"Don't be. Just take the job. Help me make this place as great as it can be." They were quiet again for a short while.

"So." She tapped her forefinger on the desk. "Practical details. For each meeting, several well-known contributors should be invited to set the level and to attract others—maybe filling half the spots. The other half should be open for submitted contributions, from younger or less well-known PIs." He nodded. She continued. "Fast turnaround. Submit the observation figure three weeks before a meeting, one week to select, two to prepare. Record comments at the meetings and transcribe right after by competent assistants. I'd need at least two."

"No problem."

"Transcripts need to be checked, again very quickly, by the contributors. We won't let them add new comments afterwards, or significantly modify what they said, but we should allow them to delete some things."

"You think?"

"Why not? If someone says something which upon further reflection is not so smart or is contradicted by facts. Think of it as their safety net. You wouldn't want people to feel inhibited about commenting on the spot."

"Good point."

The discussion continued. Ideas argued, agreed upon or given up. They pushed on. Only when Tony's PA knocked on the door to remind him about their dinner reservation did they look up. They stood and stretched.

"It's funny—" Tony said, with a smile "you actually started down this road years ago, when you were in my lab."

"I did?"

"I was teaching you how to write a paper and I explained about flow and narrative. You asked me why we needed a narrative. Like that student at your talk."

"Really?" Jessie shook her head. "Maybe this has been incubating for twenty-odd years." She laughed. "I was a bit of a show-off as a PhD student, wasn't I? Questioning everything?" He smiled fondly but didn't contradict her. "Oh, yes, and there was Cliff, my then boyfriend, the physicist. I seem to remember that what he was taught about how to publish science was different from what you taught me."

"Guilty as charged, I think." Tony shrugged. "Just passing on the lore."

"I did eventually learn how to write a proper paper," she paused, "even if it wasn't a very interesting one."

"I felt bad about that. You were the brightest of my students."

"You've had quite a few bright ones."

"Well, yes, over the years. But when I was just starting out, you were my star student. You got the most risky project."

"I wanted that project. It was the most interesting one. We used to talk and talk about it, remember? And I tried all kinds of approaches. I learned a lot."

"But the whole thing bombed. I should have foreseen that. It was my fault you didn't get anything more substantial out of your PhD."

"I got a first-author paper. From the side-project."

"You know what I mean."

"I do. But really, I. . ."

"I was afraid that impossible project made you leave active science. I still feel bad about that. You had such promise."

"I made my decisions myself." She noticed the defensive edge in her own voice, regretted it, but continued anyway. "And I've done alright, actually."

"Oh, yes, of course. I just meant. . .. No, you've done brilliantly. People really respect you."

"Well, some do, some don't. But I know them, I know the field and I have become pretty tough. Tough enough to handle a challenge like this one."

"This could be huge, Jessie."

"It could. It could change everything."

"You know, this may also help address another problem the PIs brought up: That scientific meetings have become uninteresting, a stage to perform on but little else."

"Well. . ."

"Most of what is presented at meetings is already published, or nearly so. This is fine for students and so on, but not for people who know the field. They need the new stuff. It's sad. All these bright minds in one place and no new information is presented, no substantive discussions are had."

"Except in private."

"Except in private. But what a waste! I suppose the grueling publication process is part of why people are reluctant to share work in progress. Plus simple laziness once no one expects to hear anything new." He made a face. "I'd really like to shake it up."

She smiled.

"Now I remember." She said.

"Remember what?"

"Why I chose to do my PhD with you, the youngest faculty member with the craziest projects. You were always so passionate about science."

"Now older, but still passionate."

"Something like that."

———

"Are you sure you'll be alright here? I'd see you in, but Hester sounded like she was in a bit of a panic."

He sounded apologetic, but also distracted. They had hurried out of the restaurant, not even waiting for the bill. He explained that he had a tab, so it was not a problem. But he didn't say anything further on the short drive from the restaurant back to the institute, neither about the phone call nor about the wild plans for the future that had occupied them until then. She knew there was something complicated about his home life, but not exactly what it was. They had talked for hours but neither of them had mentioned their spouses, she realized. Now was not the time to start asking.

"No problem." She pointed to the key-card in her hand. "For the main door and for room number seven as well, right? Your assistant told me my bag would be in the room."

"If Grace said that, it will be. Guaranteed."

She turned to him with a big smile.

"I'm so excited."

"So am I." He gave her a friendly, but slightly strained, smile. "And I'm so glad to have you on board. This initiative of yours—it will be a great addition to Oak Hill. Now, I hope you'll have some good discussions tomorrow morning. Come back to my office before you take off. Wrap-up and all that."

"Of course. I'll see you tomorrow." She opened the car door and stepped out into the warm and dark night. The bright lights of the parking area made the buildings almost invisible. There were a few lights on inside the main building. The annex and the guest wing, however, were totally dark.

She went around to his side. Tony's window was rolled all the way down.

"Maybe if you could wait until I've checked that I can get in?"

"Of course."

She walked quickly toward the looming, reflective building, aiming for a small metallic box lit by a faint blue light. The large glass doors obeyed her card, sliding sideways smoothly and noiselessly. At the same time, the lights came on inside. She turned around to wave at Tony, or at the car and its headlights. The car moved off.

The lobby was completely quiet. From the tour earlier, she remembered that the guest rooms were off a corridor slanting up to the right. It was dark there now. She moved briskly in the right direction, but slowed down when she heard the hard echoes of her footsteps on the stone floor. The lights of the corridor came on when she was about five feet away. Like the lobby, it was empty and waiting.

Once she opened the door marked "7", the room lit up as well. It was spacious and almost elegant in its simplicity. It had a double bed with a cheerful cover, a desk facing a wall of dark glass and a small sofa with a TV mounted on the wall opposite it. It was a perfect place to stay for a night, or for a month. Her suitcase was next to the sofa table. She saw the remote and turned on the TV for company. CNN, but she didn't mind. She sat down on the edge of the sofa, pulled out her laptop and connected to the WiFi.

She was dead tired and her brain felt fried, but once she started thinking about where she was now versus where she had been this morning, the excitement claimed her all over again. This was really happening. Her crazy, ambitious idea. She might actually do something significant here. It was incredible. There was so much to tell. But it was the middle of the night in London. Peter was asleep. She closed the laptop and looked straight ahead, smiling contentedly. The rapid flashes from the TV made no impression on her brain. She was busy savoring the moment.

———

A phone rang, in her dream. No, it was not a dream. A phone was actually ringing. She woke fully. It was Grace. She hoped Jessie had slept well. She would pick her up from her room in fifteen minutes and bring her to her first appointment. Like the day before, she seemed friendly, cheerful, very American. As am I, Jessie reminded herself. She had forgotten to set the alarm. She jumped out of bed and brought her laptop to the bathroom where she got the shower running. Dozens of emails waiting, even on her private account. She should stop giving it out—and get a new one. She ignored them, clicked create and typed pd. PDahl@codon.org.uk came up.

"Good morning/afternoon my sweets. I slept in, in my super quiet room. Had a v good chat with Tony. Tell you ALL about it soon!!! Off to talk to more people now, then to NY to meet Robert for dinner and home to you tomorrow. Love you. J."

* * *

This lobby was grand, bordering on opulent. The hotel was an indulgence, obviously, but an indulgence on Oak Hill's account. Tony had insisted. She didn't mind. It was also practical for the airport the next morning. Only after making the plans for dinner had she realized that it made it an even longer trip for Robert. She saw him noticing her and getting up from a deep leather sofa. He had changed since last time. It wasn't tiredness and strained gauntness as she had seen in Tony. It was the opposite: calm contentment. A few more gray hairs, perhaps. He moved toward her with only the slightest hint of his childhood limp. His smile was unmistakably genuine.

"Not half bad." He said, waving his hands about after the initial hug. "Someone is trying impress you, I think. The restaurant is back there." He smiled again. "But I should let you check in first."

"I'll just get the key and leave my bag." She looked at him with intent. "Robbie. Thanks for coming all this way."

"Of course. I wouldn't want to miss you. The conference is in Philly—it's not that far, really. Go on, go on." He motioned toward the reception desk. She gave him an extra hug and moved off.

Minutes later, they were being seated in the restaurant, both trying not to giggle at the maître d's formality.

"This reminds me of. . ." Jessie said.

"Yes. Those stuffy places Dad used to take us to once a year."

"I wonder why. We were just kids. We didn't enjoy it."

"At least we were well behaved kids. Once it was time to take Kathy along. . ."

"I remember that—the huge fuss over dessert. Dad was furious with her. And with Mom."

"Mom said nothing, as usual."

"She's pretty good at the silent suffering bit." She frowned. "You know, I'll never understand why she. . . never mind." She drew a deep breath. "How are they? Mom and Dad? It's been too long, I know."

"They're fine. The usual." He shrugged. "They're excited about the possibility of you moving back."

"I probably shouldn't have told them, not yet. I suppose I let it slip on purpose, though, compensation for not seeing them this time."

"It would be great to have you back, Sis." He smiled.

"Sis. You never call me Sis."

"Sorry. It's just—I don't know—middle-age sentimentality." His next smile was slightly sad. "We used to be so close. . ."

She looked straight at him and then tilted her head, quizzically. Then they noticed that their waiter was hovering. They focused on menus and ordering for few minutes.

"That's what happens." She said when they were alone again. "We grew up. And you've got Marion now." She quickly regretted the comment. It was unfair. Marion wasn't the only reason for their diminished contact. "And the kids, of course. How are they?" She continued, trying for more grace. "Happy and healthy?"

"They're good. We've..." He stopped. "No, that can wait." A pause. "It's true, you know. We'd all be very happy if you and Peter moved back. Me especially."

"Thanks." She paused. "I'll try my best."

"So will I, so will we."

"Where is this coming from? The last few times I've been over ..."

"I suppose I was too busy. Preoccupied. The academic rat race..."

"Oops, I forgot." She lit up, remembering the off-hand comment in his last email. "Congratulations, Professor. Tenured and secure. That's fantastic." She smiled.

"Thank you." He smiled back. "I am happy—and relieved, to be honest. I really felt the difference this week at the conference. The same faces, the same minor disputes, but this time I could choose who to talk to—and who *not* to talk to. Such freedom."

"I can imagine. No more worries about petty squabbles affecting some esteemed colleague's "completely professional" evaluation of you."

"It's just a few of them. Most I enjoy seeing again." He paused. "Of course, there'll be more reviews. Of my future learned contributions, if any."

"Anything new in the works?"

"Nope." He smiled. "I'm taking a break from that. Just teaching. And reading—new things, my favorite classics, anything." Another smile.

"Sounds nice."

"Oh, and speaking of Dad. Do you know what he said when I told him?"

"No." She drew out the word. Her guesses were not worth sharing.

"He said: "So, of English, is it? Well, I suppose we do need English teachers." As if... It's not a real professorship, in his eyes. The humanities..."

"Well, I think it's great, Professor." She smiled at him and then shook her head. "It's hopeless, you know. We've both disappointed him. Each in our own way."

"We are not doing so badly. Kathy didn't even graduate from that ridiculously expensive college of hers."

"He never expected anything from her." Her eyebrows lifted. "So she couldn't disappoint." She shrugged.

"Well. Who cares? Right? I happen to like my mushy humanities and my mushy professorship. And you are being courted by a prestigious research institute." He raised his glass for a toast. "To... I don't remember what it's called."

"Oak Hill Research Center."

"To Oak Hill." They toasted. "What's that about, anyway? You've got a great job already—at least you've never mentioned wanting to leave. So? Tell me."

"Well... This would be very different. There's this idea I've had for a while and Tony really likes it. It's a bit crazy. It's..."

Their starters came and were introduced in quite a bit of detail. They exchanged glances. As soon as the waiter left, Robert leaned forward.

"Tony—your former PhD supervisor?"

"The same. We've kept in touch over the years, loosely." She stopped. Although she had known he would ask, she still wasn't sure what exactly to say about the Oak Hill prospects. Ideas in her head were one thing; a job offer was another.

"You know," he said, when she had been quiet for too long, "That name, Oak Hill, and you and me, here. I'm sure it is a good omen."

"Why?"

"Remember all our tree-climbing? Up at the lake, but also at home. There was a giant tree in the back yard that we used to climb to get into the old cemetery. Wasn't that an oak?"

"I'm not sure. But it was an excellent tree." She smiled, mischievously. "I wonder whether they knew."

"Mom and Dad? That we spent our days trespassing?" He made an expression of disbelief. "Not a chance."

They laughed.

"OK. Seriously." She said. "I'll try to explain. So, this place, Oak Hill, is probably going to be *the* top place for scientific research in my area. And Tony wants me to be scientific co-director. . ."

"Wow. That sounds impressive."

"Well, co-director can mean anything." She said, but smiling. "So, this idea I had. It's about a different way of publishing science. I told Tony about it last year. Now he actually wants to make it happen." She put on a deliberate expression of disbelief. "With me in charge. At Oak Hill."

"So he wants you *and* your idea." Robert exclaimed. "Brilliant. You'll give the whole thing instant credibility. You're already at the top in science publishing."

"One of the top three journals."

"OK, so one of. But it makes sense, doesn't it? With you in charge, people are bound to take the new venture seriously."

"It doesn't work like that. I may be editor-in-chief but it's the journal's reputation that scientists care about, not mine. And this may be my idea, but it's Tony, his institute and the funding agency that will make people take it seriously." Robert was about to protest when she stopped him with a gesture. "I don't mind that, I really don't. Ideas are cheap. Reality is much more difficult. But if we can do this, it would be so. . ." She breathed in deeply, shook her head, but didn't finish the sentence.

"Well, now you've got me really curious. Explain your idea. I'll try to follow along as best I can."

She started explaining. He stopped her after a few minutes. "Wait. Narrative. What do you mean scientific papers have a narrative?"

"Well, a beginning, an end and a proper path in between. A story. A true story, of course, not fiction."

"I do understand what a narrative is."

"Sorry, professor."

"Apology accepted." He smiled. "No, I mean I don't understand how or why you use it. I thought science was all about hard facts: experiments, observations. What do you need a narrative for?"

She smiled at the echoes of her own words and recalled Tony's comments, but decided not to digress.

"It's easy to follow. The author explains what the quest is for, the overall hypothesis, for example, then how one step leads to the next, how one answer leads to the next question and so on. The story essentially recreates the discovery trail." She shook her head. "Well, it does and it doesn't. In the published story, the order of events, the order of the experiments, is often made up. Fictional."

"Fictional? No, it can't be." Robert looked aghast.

"The facts are real—don't worry—the experiments and the results. But the reasoning behind each experiment and the order they are presented in can be constructed after the fact to suit the story. Everyone does it and everyone knows this."

"I don't believe it."

"Well... Experiments should be independent of one another, so one can argue that it really doesn't matter how or in what order they are presented. Also, many findings are made by chance, not as the result of a well-reasoned hypothesis. The narrative makes it neater, more logical, more scholarly, if you will."

"Seems crazy. But I'll take your word for it."

"As I said, the main reason we use narratives is that they are easy to follow. But it is not necessarily the best way to publish science."

"OK. Got that. I guess." He shook his head. "So?"

She explained her idea as well as what she and Tony had worked out in their discussions. It took them well into the main course.

"So, this has not been done before?" Robert asked, looking skeptical.

"No."

"And it may never take off?"

"That is possible." She said calmly. He kept looking intently, and skeptically, at her face. She finally relented. "It is risky. I admit that I haven't thought too much about what happens if it doesn't work." She paused. "I suppose I can afford some risk at this stage of my life." Another pause. "You have to understand. This is a unique opportunity. If I can be the driver of something like this. . ." She realized she was waving her cutlery around and stopped, placing it instead neatly on the side of her half-full plate. She continued, more quietly. "I'm also ready for a change. I didn't realize it before this, but now I do. Eighteen years in one place is a long time, even when climbing the ladder."

"I respect that. My brave sister—not afraid of climbing, not afraid of jumping."

They were both quiet again. She continued soberly.

"Peter needs a job as well, of course."

"What does he think of your new job title?"

"Well, he doesn't know the details yet. It's all so new." Her expression clouded over for a moment. "I'll tell him tomorrow, when I get home." Another pause. "But he knows about this idea and he knows Tony had something special in mind for me at Oak Hill. And he's completely supportive of that." She locked eyes with Robert and continued. "Of course I can't ask him to give up his position at the Codon Institute unless he gets something good over here."

"That shouldn't be a problem, should it? Science in the US is very well funded. And it's the best in the world. Or so everyone tells me."

"And they are not wrong. But Peter's institute—it's quite special. . ."

"He's got a great CV, doesn't he? Dad always says so."

"Yes, he does say that, doesn't he?" She scoffed, then shook it off. "Anyway, yes, Peter has an excellent CV. He has made really important contributions and they have been properly recognized. I would guess everyone in the field knows who he is."

"So, how about the institute you are going to—Oak Hill?"

"Now, wait a minute." She laughed. "Nothing's been decided yet. I haven't even been offered the job, formally." She was still smiling as she said this. "Tony needs to take it to the board and get their approval—both for the publishing initiative and for bringing me on board. Next week I think." She widened her eyes. "It's all moving so fast. It's unreal." She took deep breath. "Maybe moving."

"Come on. You know you'll get the offer."

"Well, yes, I guess I do."

"So, there, see? And Peter could get a job at the same place, couldn't he? It's all got to do with how the brain works, right?"

"Well, yes, but. . ."

"Did you talk to Tony about Peter?"

There was a pause before Jessie answered.

"Not directly." She spoke slowly. "It wouldn't be appropriate at this stage."

"So you can ask him later."

"Yes. I'm sure it will come up. But it's complicated."

"You don't want to work in the same place? Too close?"

"No, that's not it."

"Or Peter wouldn't like it if you got him a job? He doesn't seem the type that would mind his wife being on top. I mean. . ." He glanced away briefly, as if embarrassed. "You know what I mean."

"I know what you mean." She smiled, reassuringly. "And no, he wouldn't mind. He would understand that a group leader position at Oak Hill would not be offered unless they really wanted him."

"So?"

"Tony only mentioned junior hires. There may not be anything for someone established like Peter." She said this flatly, unemotionally. Then she straightened up in her chair. "We are also getting way ahead of ourselves here."

"Sorry. Brotherly enthusiasm."

"Allowed."

The waiter interrupted them to take plates away and to ask about dessert. Jessie had not finished her oversize steak and knew she wouldn't be able to. They settled on coffee only, decaf. Robert looked around the emptying restaurant, then at his watch.

"Don't worry." He said, placing his hands on the table. "There is still plenty of time before the last train to Philly."

She leaned forward, put both her hands on his and smiled, in a warm but determined way. "Now tell me all about Nick and Bea. How are they doing? What are they doing? I haven't seen them since. . . forever."

Chapter 5

"Wow. It's perfect out there." Jessie leaned over to give him a kiss. "It looks like a sculpture. A chic modern sculpture." She smiled and tilted her head, looking outside. "And I like the way it highlights the plants. It sort of brings them inside." She gave him another kiss. "Thanks for remembering, sweetie."

"We should have thought of this years ago." Peter said, quietly returning the smile. They were well into the late-night welcome home meal. They had had some of her favorite cheeses, sourdough bread from the bakery he found last summer and a good bottle of red. He had almost forgotten about the newly installed garden lamp. It was only when Jessie described the buildings at Oak Hill and how she had returned to them in the dark that he remembered. He had gotten up to switch it on while she was talking. The effect was dramatic. "We'll get to enjoy it even more later in the year." He added with a touch of pride.

"It's perfect." She repeated and sipped the wine. "As is this." She held up the glass. "From our trip five years ago wasn't it?"

"Yes. We have another four bottles. They age beautifully." He looked at the wine glass, then at her face with a brief smile. After that, his gaze shifted back to the well-lit patio.

"Yum. . ." She said, keeping her eyes on his face. He seemed distracted, she thought. She considered whether to tell him more about Oak Hill, the beautiful setting and the pristine new labs. She had told him about her meeting with Tony and briefly about her meetings with Gustav and Junko. He didn't know them, anyway. He would know Mengyao, but she didn't mention him. To her surprise, he had not responded much to any of it, even when she managed to slip in the new and elevated job-title. It wasn't that he was negative, just disengaged or absentminded. He listened and nodded along, but asked no questions. Perhaps the conscious tempering of her enthusiasm had been too effective. Perhaps he was just tired. It was past midnight for him—and for real. She was the one off schedule, she reminded herself. She looked at him, looking out, then down, at the glass in his hands. Maybe she should simply suggest it was time for bed. Just then, he turned his head.

"And how was Robert?" He asked, unexpectedly.

"He was in fine form. It was great to see him." She smiled, remembering. "He just got promoted to full professor, with tenure."

"Good for him. We have a copy of his book, don't we?"

"He's written several, actually. We have the first one, an analysis of short stories set in rural communities, I think it was."

"Nice for Marion and the kids, I'd think. The tenure, I mean. Stability for their school years."

"They are both in high school now. Nicholas will be looking at colleges this fall."

"Already? Amazing."

"Robert asked me to give you his greetings—from Mom and Dad as well."

"Thanks—thanks. . ." He drifted off again.

"It's late, I guess." She smiled and leaned forward, stretching her hands toward his. "Maybe it's time to call it a night?"

"No, no. It's just..." He straightened up. "No. It's better we talk now. There's something I need to tell you." He took a deep breath. She waited. "There's this girl I've been seeing. Her name is Tina." He saw Jessie's eyes go wide in shock. "No, no, it's nothing like that." He added quickly. "I'd never do anything like that."

"Sorry. It's just the way you..."

"No, I'm sorry. I'm sorry for being so clumsy. It's a bizarre story, really. So this girl, Tina, is Danish. And she's a member of the animal rights group that was protesting at the Institute."

"You mentioned them. One of them came to interview you about using non-vertebrate model systems or something. Was that her?"

"It was. She had an ulterior motive, though. She wanted to talk to me, to get to know me, because..."

"Because?"

"It turns out she's my daughter." He exhaled and leaned slowly forward with his forearms on the table, hands interlocked. He looked at Jessie's face but did not say more, not immediately.

"Your daughter?" She finally echoed.

"Yes."

"What do you mean? You have a daughter? How?"

"This was before your time. She's twenty-four, so it must have happened when I was in Denmark between my PhD and my Postdoc."

"So you knew about this?"

"No, of course not. I would have told you if I had a child. If I had known, I mean." He looked at her hard as if expecting a response. She shrugged. "I had no idea." He said, with emphasis. "Absolutely none." There was anger in there, she sensed, well-contained, but still, real anger. "I don't even know who Tina's mother is."

"But that's... How can you not know?"

"She won't say."

"Come on—that's just ridiculous." Jessie shook her head. "This girl, whatever-her-name, is probably making it up. Maybe she's unbalanced. She found out that you were Danish and she got it into her head that you must be her long-lost father."

"Tina. Her name is Tina." He took another deep breath before he continued. "And no, I'm afraid she's not making it up. There was something vaguely familiar about her, so I decided I had to find out for sure."

"Familiar?"

"I'm not sure—a movement, something about her face—maybe it was nothing. But, anyway, she isn't making it up. We tested our DNA."

"You did what?" Her voice rose sharply, making it sound like an accusation. With a slight delay, she recognized that a DNA test was simply the logical thing to do. Her reaction was the illogical one. But it was too late.

"Well, why not? I needed to know." He said, defensively. "She won't say who her mother is. I had no other way of..." He stopped. She wasn't listening. She was working something out in her head.

"So you've known about this for a while?"

"No. No. I got the DNA results last night. Well, early this morning. I stayed late at the lab. I... Before that I couldn't be sure."

"But you—you suspected that she might be...? Back from when you first talked to her? That interview? And you didn't tell me?"

"She delivered this crazy bomb out of the blue. It was the day we had Hans and Alessandra and those guys over."

"That was over two weeks ago, before I left." She said, exasperated. "Why didn't you tell me? How..."

"We had people over. I couldn't exactly..."

"Not while they were here. But afterwards... I don't understand..."

"Saturday was such a busy day. You remember, don't you?" He sighed. "This was an important trip for you. I didn't want to worry you."

"Not worry me?"

"Oh, come on. Don't do that. I was trying to be considerate. You had a lot on your mind. And I really didn't know whether to take her seriously in the beginning."

"And since then?"

"Since then I've talked to her a few more times. I suggested the DNA test and... Well, anyway, the result is clear." He sighed. "I wanted to tell you in person—and only when I was sure. Now I am."

"Which test was it? How did you get it done so quickly?"

"I decided to do my own test. I figured it would be faster than a commercial paternity tests and I wanted to be sure there was no mess-up. The uncertainty was driving me nuts." She looked skeptical. He continued. "Plus, I guessed it would be easier to get her to agree that way. I could make it into a genetics tutorial, sort of a game."

"A game?"

"Well, you know..." He waved his hands about. "She's not a biologist and she had never seen a real lab before. She seemed intrigued by the place, even by the flies. So I explained how we could test our DNA, just the two of us, without involving her mother or anyone else. I also explained how genes were inherited in humans. She knew the basics, it seemed, but I wanted to be sure. Once I had explained, she immediately agreed to do it. That had to mean she wasn't trying to scam me. She wanted to know." He paused and looked at Jessie's face carefully. Her expression was still skeptical, but she didn't interrupt. "So," he continued, "we took cheek-swabs and made clean DNA preps. I showed her how to do it and we each did a set. It was almost fun, you know, getting someone interested in how..." He started to smile, but checked himself. "I designed oligos around twenty polymorphic sites. To make it easy to detect by sequencing I chose single base-pair polymorphisms. You see, in the Danish population the frequency..."

"I don't need a full list of markers." She said, with undisguised irritation.

"Well, you know how it's done." He continued. "We did a batch of sequencing runs and I sent them over yesterday. The results came back overnight. It worked beautifully, the traces were... Funny, isn't it? I hadn't done any of this for ages. Regular molecular biology, I mean. I've been doing some fly-work, but... I suppose it's like riding a bicycle. You never really forget." He caught her eye. "Anyway, one round was all it took. I'd included enough markers to be sure. She is my daughter." He frowned. "Of course you can never be one hundred percent sure. The overlap could be by chance—in theory—but statistically, it is extremely unlikely."

He waited for a response. She tilted her head slightly and frowned.

"Couldn't it be contamination?" She said. "I mean, your DNA into her sample. Your DNA must be all over the lab."

"We do genotyping of flies based on single wing preps all the time. This is no different." She was about to protest, so he went on. "Well, of course, the oligos. . . I did plenty of negative controls so I'm sure it wasn't contamination." He hesitated, then added: "She also has one copy of all my X chromosome markers, but not the Y chromosome markers. That was my other control."

"OK, OK. It get it." Jessie said, a bit sharply. Then she turned her head away and looked intently at the dramatically lit outdoor space.

Peter waited. He considered clearing the dishes, but decided against it. Instead, he sipped his wine. He glanced at her face every now and again.

"And you still don't know who the girl's mother is, even though you know pretty much when she was—well—conceived?"

"No. I've tried, but I can't think of anyone. I knew I'd be leaving soon for my postdoc in the US. I didn't have a girlfriend. I don't remember any one-night stands or whatever. Not my thing. I think you know that. So it's a mystery."

"Sperm-donor? You were at a medical school, weren't you?"

"Yes, but no. I would remember that. I've been told the experience of donating is quite strange." He tried a weak smile. "But it must be something like it. Tina's mother, whoever she is, certainly never told me she was pregnant. I would remember that, for sure. She obviously wanted to keep the baby to herself. I had no idea about all this. You have to believe me."

"I shouldn't, but I do." She paused, looking at him steadily. "You'd think it would be a hard thing to forget, so completely."

"I know, but. . ." He shrugged.

They were both quiet again.

"So what does she want? This girl—" Jessie looked at Peter again for a moment. "Tina. She's twenty-four, right? So she's an adult, more or less. Why did she contact you now? And why won't she tell you who her mother is?"

"I don't know the answers to those questions. I can try harder to find out. But I wanted to talk to you before I went any further."

"Right." She said, but nothing more.

"To be honest, I'm not at all sure what to do about this. She hasn't asked for anything."

"Hmmm." Jessie looked at him with a probing seriousness. "I don't know either. What to do, I mean." She left a long break, but kept looking at him. His expression was tense, apprehensive. "If anything." Another pause. "But you could have. . . well, never mind." She finally shook her head and continued in a clear and determined voice. "Well, so, tell me about her. What's she like?"

His expression relaxed. She could tell he was relieved.

"She's. . . well, it's strange, you know? I don't know her at all, really. But there's something—like I said before. . ."

"Familiar."

"Yes. It's fascinating, in a away, even a bit eerie."

"She reminds you of someone? Maybe her mother?"

"No, she reminds me of my own mother, I think. That's why it feels strange. It's such a young face, but still I see. . ."

He gesticulated, but it didn't help. He wasn't sure what he wanted to say.

"Is she nice? I mean, friendly, considerate, that kind of thing."

"I think so. She seems to be. She's enthusiastic, in a good way. But she's young, of course, practically a kid. It's hard to. . ."

He paused. After a while, Jessie continued. "And she seems smart? You said she wasn't a biologist, but she understood when you explained. . ."

"She knows some biology. From school, I suppose. She was studying social sciences at the University of Copenhagen for a while. I'm not sure how far she got. Then she got involved with this animal rights group, Eden." He gestured dismissively. "They're gone now. They stopped coming after the interview and all that."

"You convinced them to move on?"

"Well, Darya and I talked to Tina and we asked her to talk to them. She did. Anyway, this Eden group seems kind of low-key, unsophisticated. They just had a few people and a few posters out front. But Gerald was clearly relieved when they left."

"So she's seriously into this? The animal rights stuff?"

"It's clear that she cares about animals. The origin is sentimental, I think, childhood pets and so on. She knows about intensive farming methods, which are, well, not so nice. . . The Eden people also seem to be her friends, maybe the only friends she has here. That's really all I know. She's quite guarded, not just about her mother. And we've only met in the lab."

"Maybe you should try meeting somewhere else. The institute is sort of your castle, isn't it? It might be intimidating for a youngster."

"She doesn't seem intimidated at all. Actually, she's quite stubborn. She. . ." He stopped, nodded and said. "No, you are right. Absolutely."

"You could do something together. Find out what she likes to do and make a day of it. She might open up more that way."

"That's a great idea. I need her to tell me more about herself—and to put a stop to this mystery nonsense." He paused with a smile of relief. "You know what to do, as always." He looked at her face more carefully. He realized he couldn't read her. "Maybe together, we could. . . Would you like to. . ." He tried, tentatively.

"No." She said, with a small but determined smile and a precise shake of the head. "Maybe later on, but not now. You two have a lot to sort out first. You do that. Then we'll see."

"Right."

She couldn't tell if he was relieved or disappointed with her decision not to meet Tina as soon as possible. He wasn't sure himself.

"So you are OK with this?" He asked, softly.

"Why shouldn't I be? You haven't done anything wrong and, whatever the issues with her mother are, it only seems fair that this girl should know who her father is. She didn't make this happen."

"No, exactly. Her mother, whoever she is, created this mess." He rolled his eyes with exaggerated exasperation. Jessie didn't seem to notice. This had to be a shock for her as well, he thought. But despite that, she had ended up with a rational and empathetic response, as he knew she would. "I can't help but think Tina has been quite brave in approaching me," he continued, "and in being willing to do the DNA test. She has some gumption."

"Yes. So it seems. And you will need to get to know who she is, now that you know about her." The strain in her smile was almost imperceptible.

They got up to clear away the dishes and the leftover food. Passing each other in the narrow but familiar space, they touched carefully, reassuringly—an arm, a back, a squeeze, a quick rub. They were both tired. Going upstairs would mean sleep, surrendering to their separate dreams. They sat down at the emptied table, instead, and finished the wine. Peter touched Jessie's hand across the table. She touched his. They talked for a while about unimportant, unmemorable things. Plans for next weekend, perhaps. Well past two o'clock they finally got up again, turned off lights and ascended the creaking steps. The outdoor lamp was left on, by mistake. From this unaccustomed source, the wooden swing cast unfamiliar shadows into the quiet kitchen.

* * *

"You're back." Rissa gave her a quick hug and held on to her shoulders afterwards, at arm's length. She added a hard stare. "I was worried you'd be gone for good."

"It was just an interview. Two meetings and an interview." Jessie smiled at the welcome rebuke.

"So? How was it?"

"It was fantastic. Exciting. I. . ." This was addressed to the back of Rissa's head. She had turned away and was getting her smoothie from the bar. Her big, frizzy hair, still completely black and not quite dry, was bouncing back and forth with what Jessie guessed was flirtatious head movements. Jessie found a small table.

"I'm impressed to find you here this morning." Jessie said as Rissa sat down.

"You think I go all lazy just 'cause you're not here?" Rissa said. "Mondays are my best mornings—remember? The gallery is closed Sundays. Anyway, shouldn't you be jet-lagged?"

"I'm a bit messed up, sleep-wise, hunger-wise, but I'm not actually time-shifted. Routine is the best cure. I just need to get back to normal."

Rissa looked at Jessie, searchingly. "I can see something is wrong." She said. Jessie looked about to respond, but stopped with a shrug. She sipped her smoothie instead. "You don't have to tell me." Rissa continued. "Just know that aunt Clarissa is ready to listen whenever you need her."

"Aunt? Hardly." Jessie smiled. "We're the same age, aren't we?"

"I guess. It's just—this week. I'm hanging the new show. You remember the group show, Fine Lines and Big Form—up-and-coming female artists? They're all younger than me." She shook her head. "No, that's not quite true. Not in years, anyway. But some of them are terribly insecure. So I've taken on this agony aunt role. Listening, advising, hand-holding—sometimes literally."

"That doesn't sound like the Rissa I know."

"It isn't. It's just a blip. A necessary function until the show is up. Then they can return to being neurotic if they so wish and I'll be back in my usual, irresponsible form." She smiled, all teeth, and jiggled her many-colored bangles. "You will be at the opening, won't you? On Thursday?"

"This Thursday?" Jessie pulled out her phone and clicked a few times. "Of course. I just lost track of the days. From six?"

"Yes. But just come by whenever you're free. I need you at the end—ten-ish—to help me get the last ones out. Nothing sells after eight anyway."

"I'm looking forward to it." Jessie smiled back. "Truly."

"And your hubby?"

"For an all-women show?"

"Men are allowed in, you know."

"Well, we'll see."

The mention of Peter made Jessie think of the day he had ahead of him. How would the girl, Tina, react to the confirmation of paternity, she wondered. Relieved? Happy? Perhaps even excited? What did she really want? Jessie found it impossible to guess, just as she found it impossible to imagine being in her situation, both the growing up without knowing your father part and the finding him part. She understood that she ought to be generous. But she was finding it hard, very hard.

Soon after, she said goodbye to Rissa with a repeated promise to show up Thursday. Outside, she was met by a bright summer morning and a sidewalk full of people in a hurry. Everyone was going somewhere. She joined in. She started getting her head ready for a day back at the journal. She ran through meetings scheduled, calls she had to make. She visualized opening the door, seeing her colleagues again, her office and her desk. Everything as usual, yet everything changed. This time, coming back would feel different. Yes, she told herself, she was ready.

* * *

She looked at the temporary pass for the Codon Institute, touched the name written underneath her picture and smiled.

"Thanks."

"Not the actual truth, but. . ."

"Tina Dahl. I like it."

"Don't get too used to it. One of these days I'll get the truth out of you." He smiled, making sure she didn't take it as a threat. She smiled back. They were sitting in his office, with the door closed. "In the meantime, you should probably give me the card back."

She did, but with reluctance.

"You didn't seem surprised by the answer." He said.

"I wasn't." She paused. "But it was nice to see it like that. The facts, black-on-white. No, wrong," she smiled again, "green, yellow, blue and red on white." She pointed at the small pile of printouts lying face-up on his table. He picked them up and put them in the desk drawer. "It was kind of fun to do the lab stuff, and with my own DNA." She continued. "Like on TV, CSI or something, but for real."

"No beakers full of colorful liquids, though."

"No." She said. "A bit disappointing, that. CSI has a cooler lab."

"But seriously, you were not in doubt about me being your biological father? How could you be so sure?"

"I was sure my mother was telling the truth, when . . . when I insisted. Then you told me that you'd been in Copenhagen at the time. So. . ." She cocked her head. "And, well, somehow you just don't seem like a total stranger. Even if some of the stuff you have here *is* kind of strange." She gesticulated toward the fly room.

"It seems Mihai has you charmed. Be careful with him. Soon he will have you doing his fly-work."

"No way." She made a funny face "But he is nice, your student."

"I know. I was just teasing. I trust him with. . . well. . ."

"Does he know?"

"Who you really are? No, of course not. I haven't told anyone here. It isn't any of their business." He shook his head slightly. "I told my wife, Jessie, of course. About you, I mean."

There was a silence. Tina could not be surprised at this, he thought. He had told her about Jessie even before they did the DNA test: how long they had been married, her job, that they had no children but were very happy. Somehow it had seemed important to get all the facts straight.

"And? What did she say? Did she get mad?"

"No, no, not at all. Why would she? She was surprised, of course, but that's all. She and I met later—about twenty years ago."

"So she doesn't mind about me?"

"No. Of course not. In fact, she suggested we should do some things to get to know one another better. Non-lab things."

"Sure."

"I realized I don't even know where you live, whether you have a proper place to stay, whether you have a job. . ."

Tina narrowed her eyes. She swirled all the way around in her chair, twice, then brought it to a stop. Her voice was a bit harder now, guarded.

"You don't have to worry. I'm not looking for a handout."

"I know. Sorry. It's just . . . as long as you have a decent place to stay."

"I do. I stay with some of the other girls from Eden."

"Good. That's good." He drew a deep breath. "It's just—if we were to go somewhere, do something, it would help if I knew where you. . . London is big."

"Big is fun. So much more fun than Copenhagen." She smiled, then shrugged. "So, wherever. What should we do?"

"What would you like to do?"

She shrugged again. Luckily, he had anticipated this.

"Have you ever seen Hair—the movie?"

"The hippie one?"

"Yes. That one. It's actually great. I've never seen it on the big screen, though. I was supposed to, back when it came out but—well. . . They've got it on at the British Film Institute. What do you think?"

"Maybe." She swirled her chair again, back and forth, not all the way around. "I've heard some of the songs, I think. Not bad." Another pause. "Alright. When?"

"It's only playing a few nights. Thursday? Seven thirty?"

"OK."

"Do you know where the BFI is? Southbank?"

"I'll find it."

"We can grab some food afterwards if you'd like. There's a bistro right there." He felt the awkwardness of it all too clearly.

"OK. Thursday. Hippie-movie. Why not?" She stopped the chair again and looked at him with the beginning of a smile. "It sounds nice."

"So, time for some more flies?" He asked, with obvious relief.

"Noooo." She exclaimed. "No more flies." She tilted her head. "Tell me about when you were a student."

* * *

He had gotten there early and had picked up the tickets already. The evening was pleasant, so he took a seat outside with his Pale Ale and his phone. He noticed the raw concrete all around. Brutalist architecture and dingy old firetraps—how could they be allowed to dominate the cultural scene of a great city? A sad fact, but, for once, he did not allow himself to dwell on it. He sent Tina a message to say where he was—and, admittedly, to remind her. So far, she had remembered their appointments and had shown up on time. But this was different, was it not? He realized that he was constantly on alert for some shift in their relationship, something that he must not miss. What it would be, he was not sure. This was all so new. The mobile phone number was a positive step, he thought, as was the name of her roommate, Emily. Or was Emily the one who worked in the café? Their hangout, what was it called? Roots and Shoots. He found himself making little notes, reminders and questions. He knew a bit about the members of the Eden group now, but he had not met any of them. Apart from Alistair, early on, but that didn't count. He still knew very little about her life in Copenhagen. He did feel he was getting to know her as a person, though. The joint project last week had helped. Monday's chat had been nice, as well. He pictured her in the chair in his office, pushing herself around in circles, asking lots of questions, answering a few as well. He liked her, he realized. It was as simple as that. She was a good kid. "Kid" he thought, shaking his head. But yes. In some ways, she was still a kid. He imagined her coming along the river, dressed in one of her anti-fashion, colorful skirts for the occasion. Coming to see an old movie—to indulge him—her Dad. The word still sounded ridiculous to him.

She arrived with ten minutes to spare and a carefree smile. He got a very superficial hug. Anything else would have been weirder. He hoped the people around him realized she was his daughter, not his date. The awkwardness would fit, wouldn't it? Well, he couldn't worry about that.

"Shall we?" He said, indicating the open doors behind them.

"Yes, let's. I was just listening to the soundtrack again." She pointed to her earphones. "It's really good."

He hadn't listened to it, on purpose. He wanted it to be new again, if such a thing was ever possible. They went in.

—

"So?" He asked. "Did you like it?"

They were outside again and had ordered drinks and food. The vegan options were very slim. He had forgotten about that part.

"It was just so sad." She said, blinking rapidly. "When they were marching into that big fat airplane thing."

"The carrier. Yes, that was well done, I thought. Very effective."

"But he's going to die. You just know it. Even before the grave at the end. And he wasn't even supposed to be there. Shit." She looked up, blinking some more. "I thought it was just going to be flowers and songs and silly hippies with long hair."

"Well, there was plenty of that, too, wasn't there? The hair song—when he's dancing on the table?" He realized what he was doing but couldn't stop. "That was great, I thought. Lots of energy." She didn't respond. "Well," he tried, more honestly, "the Vietnam war was a huge thing. It defined that whole generation."

"I know that." She said, almost snappishly.

The food came. It wasn't very interesting, but eating seemed to do them some good. She cheered up. They talked about which songs they liked the best. She sang a few lines of them, softly. He was surprised and somehow touched to hear her clear singing voice. Now he was the one blinking rapidly. They laughed again. She talked about her grandparents, their craft-oriented version of a hippie lifestyle, featuring her grandmother's giant loom that produced beautiful, many-colored tapestries and her grandfather's wood-carvings. He let her talk, contentedly.

"Why was it you didn't see the movie when it came out?" She asked, a while later. "You said you hadn't."

He was surprised that she had remembered, pleasantly surprised. "My mother wanted to take me to see it for my birthday. It had just come out. I thought it sounded silly and I asked to see the most recent James Bond movie instead. I forget which one. She let me choose. I've regretted it ever since."

"But why? You could just go see it later, couldn't you?"

"It's hard to explain." He tried to smile, but what appeared was too sad to qualify as one. They sat in silence for a while.

"What is she like, your mother? My grandmother." She said "Farmor"—paternal grandmother in Danish—as if trying out the word.

He looked at her face for a long time. She said nothing. It must have been obvious to her that he was struggling. He wanted to tell her, he wanted it very badly. But he finally succeeded in not doing so.

"When you tell me about your mother," he finally said, not unkindly, "then I will tell you about mine. I promise."

"Not fair." She pouted for a while, but then let it go.

"Would you like to meet Jessie?" He said. "My wife." He added, unnecessarily.

"No." Her answer was even quicker than Jessie's. "I don't think so." She continued, more carefully. "I'm sure she is perfectly nice and everything. But I like this—just you and me—meeting, talking about stuff. Is that alright? For now? It's just. ..." Her face showed shifting emotions. Toward the end, pleading, but before that it seemed closer to fear. He was reminded that he still did not know what trauma her mother had inflicted on her.

"Of course it's alright." He said, trying his best to be reassuring. "I enjoy talking to you as well. Very much. We don't need to change anything."

Shortly after, he paid up and they said goodbye. He said he'd told Jessie he'd be back before eleven. Tina said she would make her own way home.

* * *

"Finally." Rissa said, closing and locking the gallery door. The door and the outer wall was all glass, so they could see the three women making their uneasy way down the street. They seemed to be laughing.

"Well, at least one of them should be happy. You sold both of her pieces." Jessie nodded toward the rear part of the sidewall, where two large and colorful prints were yelling for attention. The red dots were already in place.

"Xenia? Yes, I did. It surprised even me. They're a bit—well. .."

"Loud?"

"Yes." Rissa smiled.

"Xenia—that's not her real name, is it?"

"No. It's Ann-Margaret. Harrington. Not quite right for oversize canvasses of cartoon violence." She laughed, briefly. "She was one of the artists giving me hell over the hanging. She wanted the back wall." Jessie looked at the back wall. It held a collection of subtle drawings. She thought she had liked them, but she couldn't quite make them out from where she stood. She moved closer while starting to pick up glasses that had been left on the floor.

"They all call you Clarissa." She said, as they entered the small back room with their hands full. "I thought you preferred Rissa."

"I like that you call me Rissa. You are special."

"Girlfriend from another world."

"Yeah. Something like that." Rissa opened the small fridge and pealed the silvery foil disguise off a bottle of Cava. "Ta-Da!" She exclaimed. "If you have time, that is."

"Of course." They moved back to one of the front rooms. "You also promised me the tour." Jessie continued. "Anyway, it's only—" she pulled out her phone "—eleven. Peter won't be home for ages, I'm sure."

Rissa opened the bottle noisily and lost some foam before maneuvering it successfully over Jessie's glass. She filled her own glass as well. They toasted.

"So he's out with the boys?"

"Not exactly. He's. .. He's out with. .. He's out with. .. Oh, damn."

Jessie sighed and looked down, then out the window, but she didn't explain. Rissa's expression grew harder.

"He's not ... He's not having an affair, is he?"

"No, no." Jessie shook her head. "It's much worse. Sorry, no, erase that, I didn't mean that." She paused. "Peter wouldn't cheat on me. He's not like that. And, if he did, I'd be—" She gestured at the door. "Zoom. Out of here—of there."

"So?" Rissa asked. Jessie looked far away. "So?" Rissa repeated. "What's he doing? Working too much?"

"No, not tonight." Jessie grimaced. "He's out with his daughter. Bonding. Going to the movies."

"His daughter? I didn't know..."

"Neither did I. Neither did he. Or so he claims."

"But how?"

Jessie explained. She included the DNA testing, but excluded the fact that he, or they, had done it in his lab. That still felt too weird. She wasn't sure why it bothered her so much, but it did.

"So tonight they are seeing Hair." Jessie finished off, with an expression of mild incredulity, "You know, the seventies movie? Apparently it's playing at the BFI."

"It is?" Rissa smiled. "I'd love to see it again. Do you know..." She looked at Jessie's face and cut her question short. "Strange choice of movie for a twenty-something year old." She said, instead. Jessie shrugged. Rissa's expression changed to show concern. "That must have been quite a shock. For both of you." This was met by another shrug.

They sipped more Cava.

"So, the tour." Rissa finally said, refilling the glasses and putting down the bottle. She turned toward a large, fleshy nude on the wall to the right. "Powerful, no?"

Jessie didn't like it. It repelled her slightly. "It is very real," she said, "honest, I suppose, almost shocking."

"Frederica didn't want to show it, initially. I kept asking her why until she ran out of contradictory excuses. Too crude, too private, too much like Lucian Freud. Words from someone at art school she respects way too much for her own good, I suspect." Rissa paused. "Wouldn't sell, she also said. Well, it's not the easiest piece. But I've had two serious expressions of interest. So maybe."

Jessie made herself look at the larger-than-life body for a while. Then she turned to Rissa with a sigh. "You know... Sometimes when I go to museums and I see those old paintings—those voluptuous female nudes. I feel a moment of empathy, then claustrophobic panic and finally, immense relief. Back then, that was all women were. Flesh. Bodies to be desired and vessels to produce babies and heirs."

"Being desired is not bad thing, you know." Rissa smiled briefly, then turned serious. "Anyway, Frederica's nudes are very different from that. They are not voyeuristic. They are felt."

Jessie nodded. "Do you think it reflects a body-image problem?" Self-loathing, she thought, but didn't say. "The massiveness."

"Maybe, maybe not. But, importantly, it makes people think." Rissa stepped up close to the painting, as if inspecting a detail. "She's an interesting person, Frederica, and very talented. She is a big girl, but not obese, and not bothered by her size, I think. It's hard to know for sure, of course. Once we had this up on the wall, properly lit, she was happy with the decision. She allowed herself to be proud."

"I understand why." Jessie said, suddenly seeing the strength in it.

"Oh, and speaking of vessels. . ." Rissa pointed at the sculpture closest to the door. They moved towards it. The sculpture sat on a plinth. It was less than two feet tall, five-six inches wide and carved from a complex stone, mostly black but with fossils and other inlayed irregularities. The shape was smooth and smoothly curved to resemble a female torso. In the very middle was a large hole, all the way through.

"Barren, it's called."

"The stone is beautiful, the shape also." Jessie touched the cool stone very lightly, letting her hand follow its shallow curves. "It is very appealing."

"It's supposed to be dark, full of loss. But I agree with you, it is appealing. This is her other piece." She pointed at another sculpture, in a pale stone but otherwise similar in shape. "The counter-part. Bliss. Light and full of life." Jessie moved toward it and noticed the central bulge.

"A bit obvious, isn't it?" She touched it as she had the black stone. "But they are both beautiful. And tactile."

"The artist, Agatha Morris, has made several of these pairs. These two are her best ones so far, I think. Beautiful and tactile, as you say, but also very emotive." Rissa paused and caressed the smooth black stone. "Apparently, Agatha had an abortion some years ago and now she's trying to get pregnant, but failing. . She's very distressed about it. This is what one of the other artists told me. Agatha herself doesn't say much. But she is becoming a very strong artist."

They admired the sculptures in silence for a while.

"Did you ever consider it?" Rissa asked. "Having children, I mean."

Jessie straightened up quickly. Another moment passed before she answered. "Honestly, I've always known I didn't want to have children. That's why I'm so grateful I live in the present-day. We have a choice." She tipped her glass back and forth. "It's not that I hate children, I just knew that whole thing wasn't for me." She looked at Rissa. "And you?"

"I'm quite happy with my life as it is. I never felt a burning desire to change nappies and I don't go all gooey-eyed over babies. Mostly I just tell people I never met the right man. Maybe it's even the truth. It's certainly easy to say. No one questions it. They even pity me, the mothers. Poor me, at my age etc."

"That's kind of awful, isn't it? But I understand why you say it." Jessie picked up an exhibition catalogue and started leafing through it, absentmindedly. "I don't have that excuse. Most people assume that I'm a cold careerist—or that we are infertile. Luckily, they don't dare ask which." An unintended aggressiveness had snuck into her tone. She noticed it and added more softly. "No, that's not true. Not most people, just some people. Women, generally. Including members of my family." She sighed. "It seems so unfair. Men aren't labeled as cold if they don't ache for children, are they?"

"Nah, just gay." Rissa smiled as she picked up a catalogue as well. She opened it and looked at a few of the pages with satisfaction.

"It's so different for them, isn't it?" Jessie continued. "Look at Peter: Instant fatherhood. Couldn't happen to me or you."

"Nope. You'd kinda' notice. . ." Rissa laughed. She exchanged the catalogue for her glass and sipped at it. "It must be weird, though, this sudden fatherhood."

"I suppose. I probably wouldn't mind being a Dad. Maybe."

"Come again?"

"I mean, you and the one you love create this new little person together. I get the appeal of that. But without all the heavy stuff that goes with Motherhood. Mothers are supposed to love having babies—and never, ever forget. Fathers are applauded for the good they do, not scolded for what they don't do."

"That's a bit over-simplified for our day and age, don't you think?"

"Maybe, but there's still something to it. . ."

"Would you like to have a grown-up daughter all of the sudden?"

"No. I guess not." Jessie held Rissa's gaze for a while. Her face started twitching slightly. "I can't help wondering what he thinks of all the years he missed." She added, with an edge of panic in her voice. "Regret? Are they wasted years?"

"Of course not." Rissa said quickly, emphatically. "Why would he think that? You and Peter have a great marriage, a happy one. You know that. He knows that. Hell, even I know that. I've been envious for the last decade."

"I suppose you're right." Jessie picked up her glass again and twirled it slowly. "But it's not so simple any more, is it?"

"Isn't it?"

"It feels different. Like I can no longer be sure about anything." She tried to add a smile but it faltered. "Maybe I'm just not good at sharing. Silly me."

Rissa offered a gentle smile in return, but Jessie didn't see.

"You did agree about it, didn't you?" Rissa finally asked, trying to keep the tone light. "You and Peter. About not having children?"

"Of course. We talked about it, early on. Peter agreed that it was fine. Or we would never have. . . I mean that's kind of a deal-breaker, isn't it? He said it didn't matter to him, that he was happy without."

"There. You see?"

A moment later, Rissa refilled their glasses, took Jessie by the arm and led her into the second exhibition room.

"Now, let me show you something completely different. My favorite piece in the whole show." She said, with a flourish of movement that cost her some of the glass' content. "Just don't tell anyone." She switched to a tone of exaggerated conspiracy. "I'm not supposed to have favorites."

Chapter 6

"So, how is it going?"

"It?"

"You and Tina. Getting to know each other. You've had quite a few dates by now, but you haven't told me much." She finished with a brief smile, rubbing his arm from across the table. It was a Saturday morning. They had plenty of time.

"It's going fine, I suppose. I wouldn't call them dates, though." He raised his eyebrows and then shrugged. "Well, maybe. We go to the movies and we have something to eat afterwards. Or we go for walks in the park. We talk, awkwardly and carefully. There's a proximity, but it feels like it could blow up at any time."

"I'm not trying to interfere or anything. I'm just curious."

"It's two steps forward and one step back. Or two steps back. She gets defensive very easily, about her mother but also about her London life, the Eden group and what they do. And forget about plans for the future. I never get anywhere with that. So, not much progress in—what?—four-five weeks?"

"But you talk about other things, then, I suppose?"

"Well, yes. She likes going to the cinema. Thanks for suggesting that, by the way. She knows endless facts about actors that I've never even heard of. I try to pay attention, but I'm afraid it's in one ear and out the other." He paused while pouring himself some more coffee. "She has a good sense of humor, which is a relief. And she's interested in people, people around her, where they come from, their mannerisms and motivations. She's even interested in me." His smile was hesitant.

"That's only natural, I'd think. To be curious about you."

"She asks lots of questions—about my work, my years as a student, but also about my childhood and my parents. I'm afraid I insist on reciprocity there, so I don't tell her much about my mother."

"That must be difficult for you."

"It is." He looked at her quickly, not sure of the tone. She looked sincere. "She also seems to have guessed how I feel about the lab. She teased me about it once, told me to go back if I needed to."

"Perceptive."

"Yes. I think I was going on about some new results. Being over-enthusiastic."

"Hmmm"

"This was on one of our walks. We've done several of our usual walks, which has been fun. Strange, also, I mean. . . .." He frowned, but didn't finish the thought. "We've even botanized a bit, in the parks."

"The biologist at work."

"I just show her how fascinating it all is. I pick a flower and tease it apart, explaining how all the parts work. I guess I like explaining. She seems to enjoy it." He paused. "But sometimes it's frustrating. I guess kids that age, young adults, are meant to be inscrutable."

"She's about the age of your students, isn't she?"

"More or less. So maybe it's not that. I know what the students want from me. I don't know what she wants—or needs. She's also unpredictable. Sometimes she has lots of energy, asks lots of questions. The last couple of times, she's been quiet, uninterested. But I can't get her to say what is wrong." He shrugged.

"And what do you want?"

He made a sudden head movement, as if surprised. "Well, to keep contact, obviously. To make her feel she can trust me. That may take a while, it seems." He paused. "I'd like for you and her to meet." He looked over. Jessie didn't look up. "Sometime."

She poured some lukewarm coffee for herself and sipped at it.

"And you still don't know who her mother is?"

"She won't tell me. She says it would ruin things if she did. I don't see why, but I can't make her. . .."

"You must have some sort of idea by now. Really. . . You said you hadn't been a sperm donor. So there must have been an actual event. Unless you want to plead the fifth here?"

"It's not that. I know it seems ridiculous, but I'm not getting any closer. I remember what I did in the lab and writing my applications. I remember the sad room I stayed in, even, but nothing about a social life. Well, I saw a couple of the people I had done my Master's with." She gave him a questioning look. "No, they were just friends."

"Have you tried contacting them?"

"No—why?" He looked puzzled for a moment, then understood but shook his head. "No, I couldn't. It would be too weird. Plus, how would they know?"

"Tina's mother obviously remembered what happened. She told Tina who you were, didn't she?"

"She did. But that's different, isn't it? She had a daily reminder. She saw Tina grow up. She saw every little. . ." Peter paused, looking into the middle distance. He didn't notice Jessie's expression. The hurt. The hardening. "So she'd think of me—I suppose."

"Do you think about it? Miss it?"

"What?"

"The years you didn't get. Her growing up."

He bit his lower lip and looked away again. After a moment, he turned back to Jessie and said, slowly and a bit sadly:

"You can't miss what you never had, can you?"

She didn't bother to answer. It was facile, stupid. He should know better.

"What does it feel like?" She said, instead. "Being a father all of the sudden?"

"It just is. I'm still the same person I always was. Not much has changed, really." He threw up his hands in a gesture of defeat. Then he added, softly: "Well, I suppose it is different."

"Different how?" She asked, very quickly. They made eye contact, but both broke it again. "No, please," she continued, her voice neutral, "I'd like to understand."

"Well. As I said, it can be frustrating. Sometimes, I feel like telling her to go home and sort out her life and let me get on with mine. Other times, it feels very real. Important. There's a real connection."

There was a pause, a silence, longer than it should have been, but Peter did not seem to notice. "Explain the real bit." She said, her voice almost cracking.

"Explain? Well. . ." He sounded apprehensive. "Well, OK. So there's the feeling of familiarity I mentioned before."

"She looks a bit like your mother. Some gesture, you said. The DNA proved that guess right." The hostility in her voice was more obvious now. He stiffened but didn't respond immediately. She continued. "But that's not exactly a real connection. That's just genetics."

"No, it's not. Genetics is the impersonal logic of shuffling DNA around. Seeing traces of my mother means a lot more to me." His tone hardened to match hers. "You wouldn't understand."

"Don't you dare." Jessie looked furious. "It's not my fault that your mother is dead and my parents are alive."

"I didn't say it was." He waited. "But you have a choice. You can just go see them. You don't, but that's another matter."

"Sometimes we do. Anyway, we live here, they live there."

"Come on. You travel to the US for work all the time."

Peter leaned backwards, folding his arms across his chest. Jessie leaned forwards.

"Do you want my parents?" She said. "You're welcome to them. Honestly. They like you better anyway. More successful and so much nicer than their daughter."

"Whoa." His hands flew up. "Where did that come from?"

"It's true. Even Robert knows that. My father adores you."

"So he likes me, so what? That's hardly a problem." He caught her eyes and didn't let go. "Anyway, don't be so dismissive. It's not fair. You also have Robert—and Kathy." She scoffed at the latter. "I don't have any siblings."

"You father. . ."

"Half-siblings, but that's not the same. We didn't grow up together."

"You didn't see Tina growing up either."

"No, and I'm sorry I missed that."

Jessie stared at him, the words heavy in the air. He kept eye contact for while, but ended up looking away.

"Well." He finally said and looked back. "It's the truth. Or at least I think it is. You wanted me to tell you. You kept pushing. So now I'm telling you. But I'm not sure what it means, if anything."

"So." She drew a deep breath. "Tell me the truth. Are you sorry we never had children?"

"What?" He looked genuinely surprised. "I never said that."

"But are you? Is that what's behind this?"

"No. Nothing is behind this. I didn't ask for any of this. She found me."

"But now that this has happened, it's changed your. . ." Her voice was losing its strength, tears were pooling in her eyes.

"No." He said, firmly. "I love our life. And I love you just as much as I've always done." He took her hands in his.

"But would you have liked to have had children? I'm asking."

"You never wanted to and I was OK with that."

"Just OK with? You didn't agree?" She withdrew her hands.

Peter drew a deep breath. In and out, slowly, audibly. He moved backward in his chair and picked up his coffee cup. He put it back down again.

"I always thought it was up to you. If you didn't want to, that was your decision to make. It's your body. And I understood. . ."

"You understood? Understood what?"

"Look, Jessie. It doesn't matter. We are past that now. We have a good life."

"I am past that, you mean. You are not."

"Oh, come on, don't go there. You have no reason to." She didn't respond. "And you have no right to. I'm not the one who. . ." He let it hang for a moment.

"You promised." She snapped, giving him a fierce look. "You promised you'd never use that."

"Sorry, I'm just. . ." He faltered.

"Anyway" she said, sounding more normal, "what did you mean, you understood?"

"Well, I understood that with your history, getting pregnant might be too traumatic for you. Gaining weight and. . ."

"My history?"

"Your anorexia."

"My anorexia? That was long before I met you. I was a teenager, for God's sake. What's that got to do with. . ."

"You told me about it."

"I told you about it because it was part of my adolescence, that's all. Not a trivial thing, but a phase. Surely you must have realized it was long gone."

"I do. I did. It's just. . . I always assumed. . ."

"Assumed? You should have asked. If you really wanted children, we should have talked about it, properly. Jesus. . ." She let out an exaggerated, exasperated sigh. He stayed quiet. "Really," she continued, more calmly, "that had nothing to do with it. I just don't want children. Why can't anyone understand. . ."

"Anyway, it doesn't matter. I told you, I'm happy with our life. I'm sure I would have been happy with a kid or two as well. But we've both been busy. You had your career. You still have your career."

"Wow. Now we are getting somewhere." Her temporary calm had evaporated. "Now we are getting to the real truth: who to blame, what to blame. First it's my warped teenage psychology. Then it's my career. Well I'm sorry you got stuck with such an un-wifely wife who won't give you what you really want."

"Stop it, Jessie. Just stop it. You're twisting everything around. I'm not saying that I wanted to have children. I'm just trying to think it through, trying to be honest." He shook his head. "I respect your work and your career, very much. You know that."

"As long as it doesn't interfere with yours."

"What? Now wait a minute." He raised his voice a notch. "What are we really talking about here? Are we talking about careers? Yours versus mine? About why we don't have children? About Tina?"

"All of the above." She sat back and folded her arms across her chest, her expression combative.

"Look, I don't want to fight." He said. "I have never complained about our not having children and I'm not complaining now. You asked me about Tina and I answered as honestly as I could." He stopped and shook his head. "I'm not going to pretend that Tina showing up hasn't affected me or that I don't care about her. I do. I don't know her very well, but I do care. She is my daughter. It is normal to care. Is it really that hard for you to understand?"

He regretted the last words even before they compressed the air between them. They hit her face like a slap. He saw it happening, but he could not reverse it. She didn't move. He pushed back his chair, stood up and started collected the breakfast things from the table.

"We need to talk about this." She said. Her face was hard, her arms still defensively crossed.

"No, we don't. You want to pick a fight about something we decided years ago and I don't know why."

"We decided. So it wasn't just me and my warped body-image and my career?"

"Of course it wasn't just you. Now stop it, will you?"

"No."

"Yes. I can't talk to you like this. I'm going to the lab. Call me when you are ready to be reasonable again." He put the dishes in the sink and turned toward the stairs. "I'll get some stuff for dinner, if you want." He added, with forced normality.

"I want. . ." She said, inaudibly, to his retreating back. "Shit. I have no idea." She remained seated and watched him go. Her tears were flowing freely now.

* * *

From the outside, it looked exactly as he had expected, homegrown and welcoming. Yet he was not sure he wanted to go in. It felt a bit like trespassing. But it had taken him two tube lines and a long walk to get here. More importantly, he had no other clues and no one else to ask. He looked again at the exuberant sign. "Roots and Shoots" was spelled out in vegetables, the carrots very orange, the beans very green. He went in.

Two girls were manning an improvised counter at the back of the half-empty room. They were having an intense conversation, it seemed, but stopped as he approached them. He ordered coffee, black, not feeling in the mood to experiment with their smiley-faced soy milk. Remembering that he had not eaten lunch, he looked for something edible. No sandwiches, but a deflated-looking cake. "All items guaranteed vegan" it said with another smiley face. He asked for a piece of cake. He was served quickly and the girls resumed their chat.

There were two large tables in the room, one of them partially occupied by two gray-haired women with a noisy child, a grandchild he assumed, and a small, short-legged dog, like a dachshund. He walked across to the other table and saw that it was really an old wooden door, propped up on low supports. Wooden crates, inverted, were to be used as chairs. He sat down, gingerly. Next to the counter with coffee, cakes and the chatting girls, he noticed a single long shelf filled with well-used books and a few stacks of leaflets, each one a different color. On the floor below sat a row of open, wooden crates. One had a couple of decent-looking breads, one had glass jars and the rest each a few pieces of dirt-encrusted produce.

He was surprised to find so few people there. Wednesday. He was sure it was Wednesdays, the day she could never go anywhere, the one fixed point on the group's schedule. The cake was not very good, the coffee too bitter and the makeshift seat very uncomfortable. He looked at the door to the street, then at a bead curtain filling a doorway in the back corner. No movement. He picked up one of the pink leaflets lying on the table, "The arrogance of Speciesism". It was the peculiarity of the word that caught his attention. The text drew parallels between treatment of animals by humans and the most extreme forms of racism. He remembered Alistair's words outside the institute and knew that he had come to the right place. He put the leaflet down and tried the coffee again. One of the girls came over. She flashed a friendly smile.

"Would you like a pillow? We have some in the back."

"Yes, actually. I'd appreciate it. Thanks." She went through the bead curtain. He could hear a door being opened, and, for a short time, more voices. One brusque voice stood out even when muffled by the curtain. The girl came back with a red-and-orange pillow and a less-than-happy expression.

"Is the Eden group meeting here tonight?" He dared.

She held on to the pillow and eyed him suspiciously.

"Who wants to know?"

"Peter Dahl. I'm Tina's father."

She twitched slightly. "What do you want with Tina?"

"I just want to make sure she's OK. She's gone off the radar rather suddenly."

The girl still looked doubtful, but something else had crept in, possibly concern. She handed him the pillow as she spoke.

"She's not here. She didn't come last week either."

"Do you know where she might be?"

She shook her head.

"Is Emily here?"

"Might be, why?"

"I know that Tina has been staying with her sometimes. Maybe she knows where Tina is now." No answer. "Please." He added.

The girl hesitated, then turned around and went back through the bead curtain. The initial shouts of irritation were muted quickly by her closing the door. When she returned a minute or two later, she was followed by a timid-looking, dark-haired girl of about Tina's age—and by Alistair. Alistair stepped quickly past the girls and walked straight up to Peter. The girl he assumed was Emily followed him. The coffee server went back to the counter.

"I recognize you." He said, in greeting. Peter didn't respond. "Who are you looking for again?"

"You know who. Tina." Peter said, feeling old and tired even before the sparring had properly started. He stood up straight, letting his height do a bit of what his age and general demeanor could not.

"Tina who?"

"You don't use last names here. Tina told me that much."

"He doesn't know." Alistair said, now directed at Emily. "He doesn't even know her name. And he lost her again. Isn't that just hilarious?"

"Look, I'm not here to argue or to give you trouble. We are grateful you decided to end the protest at the Codon Institute. And thanks to Tina, I have come to appreciate many of the things your group stands for."

Alistair still looked skeptical, but some of the aggression was gone.

"I'd just like to find her," Peter continued, "and make sure she is alright."

Alistair sighed and turned to Emily.

"She hasn't been at my place either," she said, "for about ten days. We thought she was with you. Maybe."

"No." Peter said. "Unfortunately."

Emily looked at him, with sympathy, he sensed. Then Alistair broke in. "Look, we'll let you know if we hear from her. We know where to find you." A smirk lingered.

Peter picked up the pink leaflet and scribbled his name, email address and mobile number on the back of it. "Take this." He held it out to Emily. She glanced at it, then at Alistair. She did not take the paper. He left it on the table, instead.

"Shall we try to get some actual work done here?" Alistair said as he turned toward the back of the room. "Or just chuck it in and all head over to the lost and found?" He started walking, waving over his shoulder in Peter's general direction. Emily hesitated, but followed him.

The little dog had in the meantime moved across the room and was sniffing at something on the floor. Alistair must not have noticed because he walked right into it. He saved himself from an actual fall by a quick grab of the doorframe. The dog yelped. Emily, right behind Alistair, stepped backwards in surprise.

"Damn stupid creature." Alistair said and gave the dog a substantial kick. The dog let out a much bigger yelp and slid across a couple of floorboards before bumping into the wall.

The two women at the other table turned around at the sound. "What in the world do you think. . ." one of them said, staring at Alistair. She started to get up, but something stopped her. Emily leaped across to the dog with outstretched arms and comforting words already flowing. Alistair righted himself, parted the bead curtain and opened the door behind it without looking back. Peter could still hear the low whining of the dog mingled with Emily's soft voice as he left the Roots and Shoots café. He had placed a ten-pound note on the table as soon as he had seen Alistair. He had known that any extra time spent in that man's presence was to be avoided.

* * *

It had gotten warm again, a late summer revival. They were outside, once more in the wooden swing and the wicker-chair. The garden lamp had yet to be turned on.

"How long has it been?"

"Two weeks." He answered. "Emily, her sometimes-roommate hasn't seen her, either. She sent me an email, promising to let me know if she heard anything. She seemed nice. Emily." He sighed and went quiet.

"Maybe she's just gone back home? To Copenhagen, I mean."

"Without telling anyone? Not me, not the group, not even Emily? No. That's not like her." He shook his head. "Who am I kidding? I have no idea what's 'like her'. I know nothing."

He seemed angry, Jessie thought, and worried, of course, but mostly angry. Whether he was angry at himself for not knowing or at the world for not letting him know, she couldn't tell. But the strength of it was clear. She held on to the swing with her free hand. They both sipped at the remains of their drinks and looked straight ahead, not at each other.

"Do you think something has happened to her?" Jessie tried, anyway. "Should you contact the police?"

"And say what? My daughter is missing. Her name is Tina. Tina something. No, wait, I think her first name is Tina. I can't even be sure of that. I think she is twenty-four, blond hair, about this tall." He indicated with his hand, not realistically. "Danish. So yes, in the UK legally. But maybe she did some illegal protesting. Maybe the home office can send her threatening letters and make sure she cannot stay. If they can find her." He emptied his drink, impatiently. "No, I don't know her home address. I don't fucking know anything!"

He put the glass down, hard, the ice cube slivers defying the drama of the action. He slumped further into the chair. Jessie counted to ten before she spoke again.

"This girl, Emily. Tina only stayed with her some of the time?"

"Yes."

"So she must have had a second place to stay."

"Yes, it seems so."

"So, she's probably there, at her other place."

"Yes. But where?" He said, too loudly, too aggressively. He sighed. "Sorry. Damn it."

"Could we maybe. . ."

"Look, you don't have to do this." He said, in a tone that was neither conciliatory nor particularly pleasant. "This has nothing to do with you." He paused, then added almost bitterly: "Just leave it." He looked away.

Jessie drew a quick breath. She let the air back out, slowly. A dozen answers, hurt, angry, furious, sarcastic and sad, crowded her mind. She shut her mouth, hard, sealing it. As she had done the last couple of times they had had this conversation— or a conversation much like it. Carefully, she steadied the swing. Then she slid forward, not looking at Peter, and stood up. She went inside to prepare the salad.

Ten minutes later, Peter came inside as well. She was still chopping the greens. He started laying the table for them on the counter. He found the bottle of white she had put in the fridge last night and opened it. He offered her a glass.

"I'm sorry." He said. "I didn't mean it like that. It's just. . .."

"I know." She took the glass with an understanding half-smile, but an impersonal one. "You are worried. And you feel powerless, which makes it worse."

He shrugged and slid onto one of the high chairs. Jessie placed the bread and the salad between their two plates and stepped up onto the second high chair. She looked at him for a short while. Then she served herself and started to eat, silently. He did the same. They were both facing the kitchen area.

"I'm going to Oak Hill on Monday."

"Another interview?"

"No. The board has already approved the new initiative. And me. I told you all that, last week. It's going ahead."

"So you are... you've said yes?" He stopped eating and turned to look at her directly. She didn't turn, but after a while she spoke, quite softly, into the empty space in front of her.

"I've... It'll be a trial period, for now. Tony wanted me there for the junior investigators interviews, so....."

"But we haven't decided. We..."

"No, but we've discussed it. You said, quite clearly, that you were supportive and that I should go for it, if..."

"But that was before, before..."

She turned and looked at him, steadily, waiting for the words. She both feared them and wanted them. But they didn't come.

"I know." She said, instead. "For now, I'll just be trying it out. We need to get the new initiative rolling. Tony would like it to..."

"But you could do that from here."

"No, it has to be there. I explained all that. The meetings, remember?"

"Yes, but for getting it started, planning, inviting people and so on, you don't need to be there."

"We've already done most of that. By email and Skype."

Surprise was all over his face but she did not see it.

"And the journal." He said. "You can't just leave them all of the sudden. You have a great job there. People would kill to get that job. I mean... Don't just throw it all away. Think about it."

"I have." Her voice was firm, her expression steely. One of the many things she had learned on the job was making a decision and sticking to it. "My deputy is more than happy to step into my shoes. He couldn't hide his excitement when I told him this morning. He kept saying how excited he was for me, but I understood. It's all fine."

Peter had put down his cutlery. He didn't even pick up his glass. He just stared at Jessie, as if he didn't know who he was looking at.

"But what about me? What about us? You can't just leave. Next week. Damn."

"I am not leaving you. You are my husband and I love you. You know that." She gave him a slightly forced smile and put her hand on his lower arm. "I'll just be working over there for a while. If everything works out, you'll join me later. We can go back and forth in between. The flight is not that bad, it's only....."

"But this is a big deal. Moving to the US. It's not just something..."

"I know it will take a while for you to get everything settled. You'll need to find a job you're happy with and there's your lab to deal with: who can move, who can't. It will take some time to work that out. Plus you have other things to sort out here."

She looked at him calmly. He looked back, searchingly.

"But I can't be sure I'll even get a job there."

"Of course you will. With your CV, all the new stuff coming along..."

"What did Tony say, exactly? Did he promise me a position? My lab is small, so that could work. Eventually. I'd just need a bit of time to. . ."

Jessie swirled her glass and bit her lip. She looked out the window. The evening was closing in.

"Oak Hill is only doing junior hires for now. That's all I know."

"For now? All you know?" He was aware of the desperation in his voice. He wished it wasn't there. But he needed to say this out loud. "That could mean there won't ever be a job for me at Oak Hill."

She didn't answer immediately. When she did, her tone was gentler than before.

"There are many other places in the region, great universities, great departments. They'd be happy to have you."

"I don't have any US grants—and no history of getting them."

"You'll get a generous start-up package. You know that."

He looked straight ahead, not answering.

"Once the mouse numbskull paper is out," she continued, "everyone will be reminded of how powerful your genetic approach is. You told me. . ."

"That we were trying." He said and sighed. "Ilana's first results are not looking promising."

"I thought. . ."

"So did I. But you have to do the experiment."

"The mutant mouse has no phenotype?"

"Well, we don't know for sure yet. She is repeating the experiments and doing some additional tests. But it looks like it's not as straightforward as I had hoped."

"So it'll take a bit longer. You'll figure it out." She said, calmly, reassuringly. "And you'll find new stuff. You always do."

They looked at each other. Both sets of eyes were flickering, searching.

"The thing is—I can't move right now." Peter finally said. "I can't even go on the job market. There's too much going on here."

"The lab and. . ."

He shrugged.

"Yes. Tina. I can't just disappear as well. I have to be here. You have to understand that."

"I do." Jessie said, quietly. It was better that he admitted it, she thought. In a way. In a way it was worse. He might have given in. He didn't.

"Can't you postpone?" He asked.

"No. I promised Tony I'd help with. . . I promised that I'd . . ." She looked away, then back. "And I need to—I need to do this."

"OK."

"OK?"

"OK."

Chapter 7

With all the coming and going, Kastrup airport was pleasantly busy. Some adver-
tisements were in Danish, others in English, he noticed. He liked the dark brown
wooden floors. Welcoming, yet practical. Overall, the architecture and style of the
place was attractive; it felt comfortable and only vaguely familiar. These immediate,
positive impressions calmed him. This time it would be different. He was different.
He also had a good reason to be in Denmark. He had a mission. He had two days and
a limited number of clues. It felt good to be doing this, finally.

His passport was given a quick glance. A bored nod directed him through. He
went straight to the Metro. The Metro had been new to him, last time. So much had
been new, but he had not been able to take it in. In a strange sort of panic, he had
gone straight to the conference venue and straight back afterwards. He had spoken
English to everyone and had not revealed any knowledge of the city. His name and
his affiliation had made the deception possible. Probably no one noticed and
certainly no one would have cared. But he felt like a fraud afterwards and had
declined subsequent invitations. He never told Jessie about it.

Peter did not want to think about Jessie right now. She was doing what she had to
do. He was doing what he had to do. That was all.

The doors closed and the compact little train started its driverless trip into the city.
They passed some new buildings, a stretch of red brick suburbia and picturesque
allotments before heading underground. A couple of stops later, he alighted at
Nørreport. The station was crowded but not too hectic. It was Saturday, after all.
No one pushed or made a sour face. A foreign-looking youth helped a frazzled thirty-
something Dane get her baby carriage on the train. Her expression switched from
initial worry to an exaggerated smile. The skylight over the escalators held the
promise of a beautiful day. He surprised himself—he smiled.

At street level, he looked briefly toward the familiar pedestrian streets before
walking off in the opposite direction. He noticed that two large, airy market halls had
replaced the grubby outdoor vegetable stalls of his youth. He approved. Copenhagen
would never be truly grand, he thought, and it would never be the swirling mix of
humanity and languages that was London. But it was not bad.

The hotel fit in nicely with its neighbors: five stories tall, brickwork painted white
with semi-ornate moldings on the lower floors and windows made up of multiple
smaller panes. A hundred years plus a bit, he guessed. The entrance was on the
corner, into a low-key reception area that was continuous with the hotel bar along
one side. He had chosen this place for its convenient location and moderate price.
The two-star hostels available at half the price were just too depressing.

"Mr. Dahl." She took his credit card with a friendly smile and found his booking
quickly. "So you are staying with us for two nights?" To his surprise, she was
speaking English to him. It was logical, of course, with a card from a UK bank, a
booking from abroad and his name.

"Yes." He answered, in Danish. "Just two nights. Visiting family." He had no
idea why he had said that. She smiled pleasantly but did not give any indication that

she cared, neither about language nor his purpose. Why would she? She told him that the room was ready, gave him his key, pointed to the elevators and mentioned their happy hour with a free glass in the bar at five. Then she turned to the person who had come in behind him.

Ten minutes later, he was back on the street, phone in hand for checking the map if needed. Tina's many descriptions had given him enough, he thought. Her grandparents' place had to be central, as Tina and her mother travelled there frequently, by bicycle or on foot, from an apartment not far from the central hospital. There had to be space for her grandmother's loom and the staircase was one of Tina's favorite places to sit—so maybe a small house. There was at least one good-size rabbit pen and some green common space, where she had played with a changing assortment of children. Various unusual adult characters had also featured in her stories; she referred to them as hippy-types. From all this, he had guessed either Christiania or Brumleby. He had been to Christiana, the "free town" of Copenhagen, a couple of times in his teens, on hash-buying trips with friends from school. For some reason, he felt less comfortable going there now. Besides, Brumleby was closer.

He walked along two of the city lakes and turned left at the bottom. He didn't need the map, he realized. The commercial street, Østerbrogade, was noisy, full of buses, cars, people and lots of bicycles. As at the airport, he was struck by the easy familiarity of it all. But without the anonymity of international travel he also felt exposed. The mixture of closeness and distance, of known and unknown, was disconcerting. He felt slightly nauseous. Nervous, he guessed. He pulled his mind off introspection and forced himself to look closely at the buildings he passed.

Five minutes later, he reached his destination. The rows of nearly identical yellow and white terraced houses were charming, with a few trees, well-trimmed grass and occasional benches out front. Despite the nearby traffic and a vague suggestion of army barracks, he felt a peacefulness descend. He walked down one of the little streets and came across two older men sitting in the sunshine. He introduced himself and asked if they knew of a couple in their mid- to late seventies with a granddaughter named Tina. He was referred to similarly old, but spry-looking woman, who was standing by one of the trees in deep discussion with a small child. Gertrud, he was told, knows everyone. The child seemed to be wielding a toy shovel as a sword. He waited until Gertrud had convinced the child that battering the tree was not a good idea before he walked over.

"A grandchild named Tina, you say? But you don't know their names?"

"Tina has disappeared and frankly, I'm worried about her. She talked a lot about her grandparents, Mormor and Morfar, so I thought I'd ask them if they had any news. Mormor has a big loom in the living room that she uses a lot, if that helps. And they keep rabbits. At least they did fifteen years ago."

"You must be talking about Anna and Svend Søndergård. Her tapestries are quite famous."

"Do they still live here?"

He was in luck. They did. Gertrud gave him directions for the house, but then went with him anyway. The house looked like the others, but the loom was visible from the outside, standing near the front windows at a slight angle. A head of unruly gray hair was visible as well, bent over something attached to the many treads. Gertrud knocked on the window and waved as the woman, who he assumed must be Anna Søndergård, looked up. They met at the front door.

"This man seems to know your granddaughter, Tina." Gertrud said by way of introduction. "I thought you'd want to talk to him." Anna's eyes leapt quickly to Peter's face and sparkled with what he thought was hope or joy. Her eyes were green with flecks of brown. They rapidly developed a questioning and intense stare, as if he were a puzzle to work out or a tapestry to perfect. The rest of her face was a strong and weathered type of old: well wrinkled and well tanned with almost invisible lashes and eyebrows. Mid-seventies seemed about right.

"Do you know where our Tina is?" She said.

"Not at this moment. I'm actually trying to find her." Peter said. "But I've been in contact with her recently, up until a couple of weeks ago. In London."

Anna looked pleased, then worried, then pleased again.

"Svend." She called, rather loudly, to a person bent over in a nearby vegetable patch. She walked a few steps toward him. Although she looked quite fit, she moved stiffly, perhaps from the many hours at the loom. "Svend. Come here." He straightened up, caught the tone of urgency and started moving toward the house. In the meantime, Anna turned back and Gertrud said her goodbyes.

"Can we go inside, perhaps?" Peter asked, when he had been introduced to Svend. Svend had coloring and weathering remarkably similar to his wife's, but clear blue eyes. "There is a lot I'd like to tell you." He paused and smiled. "Tina talks about you all the time. She clearly loved coming here as a child. You two, the cozy house, the rabbits." Anna's face showed this was a welcome comment. He was asked inside.

They went past the loom into the largish back room, which seemed to be a combined kitchen and eating area. Its centerpiece was a large, rectangular table made of rough-looking but smooth-feeling pale wood. It was also the designated spot for old newspapers, old mail and much else. One of the white-painted walls was completely covered with drawings made by one or more children. He did not have time to study this cheery collection in detail. Not yet. Coffee was offered and accepted. He and Svend sat down, but didn't start talking until Anna joined them.

He started explaining. He told them about his job and about his initial meeting with Tina in slightly vague terms, emphasizing her love of animals. He mentioned her unexpected and confident belief that he was her biological father. He talked about the time they had spent together, at the movies, in the parks. He tried to paint a picture they could recognize and be comfortable with before he had to admit that he didn't know who Tina's mother was.

"So Tina didn't tell you about her family?" Anna asked. A worried frown had replaced the eager smile. "Then how did you find us?"

"She told me lots of stories. Clearly, coming here as a child meant a lot to her: you two, the rabbits, her special place underneath the loom and other favorite hide-aways." They nodded. "But she never told me her last name or her mother's name. She wouldn't explain why she was being so secretive. All she would say was that she wanted me to know her "for herself.""

"Then how can you be sure that you are her father?" Anna asked.

"We did a DNA test."

"Yes, of course." They both nodded.

"How did she find you? Do you know?" Svend asked. He hadn't said much but was paying close attention.

"Apparently her mother told her who I was." Svend and Anna exchanged puzzled looks. "Quite recently, I think. Tina said she had insisted on being told and her mother had finally given in."

"That is a surprise to us." Anna said, after a pause wherein she had shared a longer look with Svend. He had finally nodded as if to indicate for her to go ahead. "You see, Maj, our daughter, Tina's mother, told us that she didn't know who Tina's father was. We figured that either she really didn't know, which seemed unlikely for someone as responsible as her—but not impossible, of course—or she might be trying to protect Tina somehow." Anna stopped and searched Peter's face for a reaction. He didn't have one. He shrugged, lifting up his hands. "Are you married?" Anna continued.

"Yes, happily married for eighteen years now." Anna's steady gaze on him remained unchanged, so he continued. "I wasn't married back then, or even in a stable relationship. I was a hardworking graduate student, not violent, nor a drunk, a gambler or a member of an extremist group. I'm still not. I'm quite ordinary, really." Anna gave him a slightly apologetic smile. "I don't have any children. Other than Tina, that is." He paused. "That I am aware of." He finished with a weak smile.

"Well, we didn't know the details and we didn't want to pressure her." Anna continued. "We just loved little Tina from the start." Svend nodded.

"I understand that." Peter said, somewhat vaguely.

"So she knew all along—and told Tina recently. How odd." Anna glanced at Svend again. "Even more surprising is Tina not telling you about Maj. Why wouldn't she? They've always been so. . . And you never knew?"

"I had no idea. Even getting a name, Maj Søndergård, doesn't make it any clearer to me. I assume she is a doctor? Tina sort of indicated that."

"Yes, she is—a dermatologist. She chose a specialty with decent hours, so she would have time to take care of her daughter. Maj was a wonderful mother to Tina, all through those busy years. We've been so impressed. She didn't have much time for fun or for going out, but always time for Tina. Isn't that true, Svend?" Svend murmured his assent. "We never knew what happened, exactly, when Tina suddenly left. I mean, once the teenage years came along, they had their ups and downs and their arguments. That's just the way it is with mothers and daughters. But it's been such a long time now. Almost two whole years." Anna stopped talking and looked off into the distance. Svend poured some more coffee but didn't take over.

"At first Maj didn't admit anything was wrong," Anna continued "but when we didn't see Tina for months on end and when she didn't answer her phone—well, it was obvious that something was amiss. Finally Maj told us that they had had an argument, a serious one. Afterwards, Tina had left with a packed suitcase, slamming the door on the way out. She had said she was going abroad and not to try to find her. So we didn't involve the police. Maj said that the lack of contact was Tina's way of punishing her. She wouldn't say what the argument was about, though." Anna fell silent again. "But she was worried, of course she was. As were we. But then the postcards started coming and it got a bit easier to bear."

"Postcards?"

"Yes, Tina has been sending us little postcards every couple of months, telling us she is fine, not to worry about her. Always an animal card, of course, and postmarked central London. She must have been there for well over a year."

"Really? I didn't... Well, never mind."

"The last postcard was early June, so we were starting to worry again. We're relieved to hear that she was just busy...." She shook her head. "You will find her again." She added, with unexpected certainty. Peter didn't feel quite so sure, but much better now that he had more information. He decided against telling them about Tina's mood in the week before she disappeared. He didn't know what it was about, anyway.

"Do you think I could talk to Maj?" Peter finally said. "I'd like to. If she wants to talk to me, that is."

Anna and Svend exchanged looks again.

"We think that you should." Anna said, pursing her lips slightly. "We will tell her what you have told us, naturally. But it seems only reasonable that the two of you talk directly." She rummaged in the layers of papers and pulled out a pen and an empty envelope. She scribbled two phone numbers on it.

"This is Maj's number. She lives quite close to here, but why don't you call her first?" She smiled, cautiously. "The number below is ours. Just in case."

"I will." Peter took the envelope and glanced at it before putting it in his pocket. "And thank you so much. It's been so nice to meet you. Finally." He got up and shook hands with both of them, in turn. Svend gave him a two-handed shake and looked Peter in the eyes, kindly. Peter understood why Tina had felt so comfortable around these two. "May I come back afterwards?" He asked. "To hear more about Tina as a child?" He had a sudden inspiration. "If you have any pictures, I'd love to see them. I feel that I've missed so much."

"Yes, of course." Anna said, this time with a big and generous smile. "I will find some pictures for you."

———

He called Maj an hour after he had left Anna and Svend. Finding them had gone so fast and he needed some time to process the visit. He also needed to prepare himself for the next step, which, he realized, might be momentous. A loop around the lakes seemed a reasonable intermission. He walked fast, not really looking at anything, just thinking. He thought about Maj, whether he would recognize her once he saw her and how she might react to him. He thought about Tina and her grandparents. She had described them in such great detail. Had she been leaving him clues on purpose? But then why the secrecy? So he would have to visit? Most likely it wasn't quite so calculated.

Maj had heard the news from Anna and Svend already. The conversation was short. He got her address and an invitation to visit that afternoon at four. She had her weekly shopping to do beforehand. He was content to have some more time to himself first and only briefly wondered why she bothered to explain the timing. The voice over the phone was neutral, impossible to decode. Turning back toward the city center, his first thought was of a nostalgia walk through the old pedestrian streets. The weather was cooperating and he was in the mood for it now. His second thought was of food. It called him into a big, busy bakery. He was hungry and ate the expensive, but excellent, sandwich standing near the window. Strangers kept him company, each for a minute or two: Kids in strollers, reaching toward parents for their treats, running groups done for the day and reaping their rewards, girlfriends giggling while looking at their phones. Ordinary life, flowing by. He enjoyed it. Afterwards, he walked to the bottom of the pedestrian street and found a café on the square. It was the same one, even after all these years. He took it as a good sign and sat for a coffee in the sun. He watched people and buildings and walked some more. He had decided not to worry about Maj and, surprisingly, his mind obeyed.

—

Two rows of five buttons, with a slot for a name next to each of them. The top ones had paper stickers, several layers thick, the others had properly fitted labels behind the protective glass. Some names were typed. Søndergård was written by hand, in a clear, confident script. He pushed the button. The door buzzed a couple of seconds later. She was expecting him.

"It's on the third floor, the door to the right." She had told him. This door was slightly ajar when he got to the landing. Informality? Trust? A challenge? He didn't know. He pushed the door open.

"Hello?" He called.

"I'm in the kitchen." A woman's voice called out. "To your right."

The narrow hallway had a slightly rough, wooden floor, white walls hung with black-and-white photos and doors of pale wood, all but one open. Near the entrance, to the left, was a longish rack heavily laden with coats, scarves and bags. They made him apprehensive, all these personal items, but he forced himself to step further inside. Most of the items were in muted colors, with two or three cheerful pieces standing out. He wondered briefly if those were Tina's. Then he stepped into the kitchen. This room was quite narrow as well, but cozy, lived-in. At the far end was a tiny café-style table in front of slender glass doors to the outside. Between him and the table stood a woman of about his age, looking at him with unnerving directness. She had chin-length blond hair, possibly dyed, and wore a loose cream shirt, matching pants and sensible, brown shoes, presently planted well apart. She was holding on to a kettle on the stove. The water was boiling. She took her hand off the kettle and offered it to him as he stepped forward.

"Maj Søndergård. Good to meet you." She hesitated. "Again."

"Peter Dahl." He shook the overheated hand and looked at her face more closely. He had imagined endless times what might happen at this moment: He would recognize her and the past would come flying back. Or she would erupt in anger, accusations and bitterness. Both now seemed unlikely. She remained calm. His memory did not come to the rescue. All he could recognize, or thought he could recognize, were hints of Tina and possibly of the older faces from earlier in the day. He had to go the slow route.

"So you heard about me from Svend and Anna?" He said, as they sat down at the small table. The mouth-watering aroma of a homemade cake just out of the oven had made it impossible to refuse yet another cup of coffee. The doors were partly open to the small street below. South facing, it seemed. He got the sunny side.

"Yes." She looked at him, searchingly. "This does feel rather strange. After all these years." Perhaps she didn't remember him either, he thought. "But I knew that Christina was in London, from the postcards."

"Christina?"

"Yes, my daughter." She paused, her eyes still on his face. "I know my parents call her Tina, but her name is Christina. She didn't tell you?"

"She told me very little. Not her full name and next to nothing about you, except that you were a doctor and lived in Copenhagen. She did say the two of you weren't on good terms right now, but she wouldn't say why." He paused, hoping for clarification. He got only a continuation of Maj's guarded inspection, which was making him a bit uncomfortable. "Tina contacted me two-three months ago." He continued. "She was part of an animal rights group protesting at the institute where I work. We spoke about animals and research, and then, out of the blue, she said that I was her father. I was more than a bit surprised. Shocked, actually." Maj smiled faintly, knowingly, into the silence and sipped some coffee. Noise from the street below filtered in.

"How is she?" Maj asked. "Does she have friends over there?"

"Well, the animal rights people. That's all I know about. But they are. . ." Peter's voice faltered. He looked down, then back up at Maj's face. He took a deep breath. "I'm sorry," he said, "but I don't recognize you and I don't remember what happened back then. I don't mean to sound callous, but twenty-five years is a long time." He paused, giving her space, but she declined it. "You didn't tell me that I was about to become a father, did you? I would have remembered that." A slight twitch on her face, but she didn't answer. "Anyway, Tina agreed that we do a DNA test. So I know she's right; I am her biological father." Maj looked unsurprised. "I'm a geneticist. I have a lab at the Codon Institute in London and we did the analysis there. Just the two of us, some PCR reactions and the faithful sequencing machine. An odd sort of bonding, I suppose, but. . .."

"I know." She said.

"About the test? She told you?"

"No, Christina hasn't spoken to me for a while, as you know. We had an argument and she decided to punish me with silence. I didn't think it would last this long, but. . . No, I meant I know about you. I've kept tabs on you, over the years. I know about your research career, in the US and the UK. You study short-term and long-term memory, right? Something about stress-induced forgetting?"

"More like reward-induced learning, but. . ."

"Your papers are impressive. I admit I don't get all the details, but I've certainly heard of the journals. Top journals. So. Memory. This was what you always wanted to study, wasn't it?"

"Yes, it was. But. . ." He faltered. "So we really did know each other twenty-five years ago?"

"We did."

"Look, I'm sorry, but please tell me what happened. I want to understand. And I'd like to be there for Tina in some way. I know it's late in the game, but I've spent time with her now and I feel. . ."

"That you know her? That you are close?" The sarcasm was not heavy, but definitely there. He didn't answer. As she went on, the tone became harsher. "I understand from my parents that she has left you as well. She has moved on. That's why you are here, isn't it? Because she has suddenly terminated all contact with you and you don't understand why. Maybe she just doesn't need you any more." She looked at him with a strange mixture of contempt and reluctant sympathy. Then she looked back down, focusing on her coffee cup.

"Yes, I suppose it is." He said, slowly. "I thought she might have gone back to Copenhagen. Or at least made contact. Maybe she had contact with you the whole time. I never knew how much of her story was true."

"What story?" Her eyes darted back up to his face with a hint of panic.

"Just that she had grown up in Copenhagen, had had a pretty happy childhood, that there was no stepfather, that you were not currently in touch, that it was her choice, unexplained," he moved the fingers of this right hand one by one, as if checking items off on an internal list "and that you had told her who I was very recently. From stories of her childhood, I learned about her grandparents, which, ultimately, was the only way I could trace her." Maj nodded, visibly calmed. "So I had to come here. I had to try." He sighed. "Please tell me what happened between us. I'd like to know."

Maj broke off some cake from the slice on her plate and ate it slowly with her face turned toward the open glass doors. There was a minor commotion on the street below. She seemed not to notice, lost in thought. Finally, she turned and caught Peter's gaze. He saw that her eyes were like her mother's, not pale like Tina's. Her face showed no anger now; instead, it suggested kindness, perhaps nostalgia. He could suddenly imagine the much younger face—a face that was smiling, laughing, pretty and fresh.

"I guess I'd better tell you, then." She said. "I was a medical student. We had our biochemistry practicals at Panum. The course was run by Torben Petersen."

"I was back in his lab for a few months."

"Yes. You had just finished your PhD work in Cambridge, waiting to defend your thesis. I guess you had to make some money, so you worked as Petersen's teaching assistant." He nodded. "You made sure we all knew where you had done your PhD and that you were going places after this unfortunate, but at least temporary, dull drudgery of teaching medical students."

"Ouch. Was I that bad? Really?"

"Pretty bad. But you were young, I suppose. So in hindsight. . . Well, I was even younger. I thought you were a total asshole. Not bad-looking, but an asshole."

"Thanks."

"Well, you were. Arrogant, like I don't know what." She showed a hint of a smile. "Then you crashed our Christmas party. I suppose we may have invited you. We all had a bit of a crush on you. I was drunk; you were drunk. I succumbed to your charms and invited you to my place."

"You remember this?"

"Of course I do. I wasn't that drunk. You were a good catch, in a way." This time she really did smile. "And there were consequences, to remind me."

"So I didn't do anything terrible?"

"You didn't force me or trick me. I was no innocent—not terribly experienced but also not unaware. I didn't have a boyfriend and I wasn't on the pill. I remember deciding not to care. I remember because it was a dreamy, silly, stupid decision, but it was an actual decision."

"And when you found that you were pregnant? Then what?"

"By then it had become clear that it was a one-night stand. The course was over, exams also. We were in different spheres." She looked at her coffee cup, not at him. "I decided not to bother you about it."

"Not to bother me?" He couldn't keep the disbelief from his voice. "But you knew that it was my child? For sure?"

"Had to be. There were no other possibilities." She paused. "I heard from someone that you had gotten your postdoc fellowship and you were moving to the US. Maybe you had already moved." Another pause. "Deciding to keep the baby came easily. I knew that I didn't want an abortion. You were reasonably attractive and obviously smart. I could have done worse. To simplify things, I chose not to tell you—or anyone."

"Isn't that a bit selfish? I mean. . ."

"Selfish? Seriously?" Her apparent calm turned to anger in a flash. He flinched, causing the flimsy chair to tilt backwards, dangerously. She continued. "Do you think it was easy, raising a kid on my own, while completing my medical studies? It was not, I can tell you that. It was hard work."

"I believe you." He forced himself to toughen up. "But you made the decision, on your own. Isn't having a child supposed to be a joint decision?"

"It was my body, my baby and my decision. You wouldn't have wanted to be weighed down by a child. That was absolutely, completely obvious. You had big plans. You would have tried to talk me into getting an abortion. Just like my parents." He looked surprised. She caught it. "Oh yes, they did. Gentle, earthy, free-living Anna and Svend." Her tone was sarcastic, but with feeling, with hurt. "They tried quite hard to convince me to get rid of it, get rid of her—"for my own good". I hated them then."

"But they clearly love Tina. And they seem to have cared for her quite a lot over the years."

"Of course. Once Christina was there, the cutest little baby with ten little fingers and ten little toes, they begged to see her and to babysit her."

"It must have been nice to have them so close by."

"I needed the help, I'll admit that. And they were there for me, for us." She paused, remembering something perhaps, and added a wry smile. "But then it's easy to be the grandparents, isn't it? Theirs was the place where everything was allowed, all fun and games. Home was less exciting."

Peter was reminded of Tina's stories about hiding so she didn't have to leave her grandparents' house. He tried to imagine being the hard-working single parent looking for her, calling impatiently, a hundred and one chores waiting at home. He didn't succeed.

"You could have told me that you were pregnant and that I was the father, but that you wanted to go solo. I would have respected that."

"Maybe." She said, somewhat ambiguously. "Anyway, by not naming you, I saved you the worry and the child-support. I raised a beautiful, happy girl all by myself. I'm sorry, but I really don't see why you should be complaining."

"I'm not complaining. It's just... It takes some getting used to. And she may not be happy now. We are discussing a young woman, out there somewhere on her own, who has now chosen not to be in contact with either of her parents."

"Parents? I don't think you qualify..." She said quickly, but stopped. "Well, yes, right now she's protesting. She's always been very passionate. When she was a child, she was so happy. She radiated happiness, singing and jumping about, always playing silly games." She paused, smiling. "She loved baking. The two of us would stand over there," she pointed to the long counter, "and make all kinds of treats. For the whole afternoon, sometimes, on weekends." Another pause, the smile receding. "But the tantrums... She reacts strongly. Always has. Right now, she needs to show her independence. But she'll come around. We'll be fine again." Maj didn't look so sure, Peter thought. She looked sad.

They were quiet for a while. He ate some cake, without really noticing. The sun was still shining, but had moved away from his seat.

"We went to see Hair, Tina and I." He finally said. "You know, the movie, the musical? She loved it."

"She did?" Maj said, her voice light, hopeful. He was inordinately pleased to have put even a small smile on her face.

"She did. And she sang little bits of it afterwards. She has a really nice singing voice." He shook his head. "I have no musical abilities, at all."

"She does, doesn't she?" Another small smile appeared. "Me neither. No musical talent. My father has, though. He plays the guitar quite well." She sighed. "Hair. I'm not surprised that she liked it. Everything to do with my mother and father and their cottage-style, naive approach to this world attracts her."

"Is that what started her on this animal rights interest, do you think? It seems to have taken over her life somewhat."

"It has, hasn't it?" She nodded, frowning slightly. "She has always loved animals. Or never grew out of it, more accurately. There was a whole gang of girls like that, around fifth grade. They all became vegetarians, too. Later on, they had some school project about farming and Christina turned full-blown vegan. She's never relented. The others did, but not her. Strength of character." Peter joined her in a reluctant half-smile. "She's very healthy, knows all about protein and nutrients, so I've stopped worrying about that. But the cake-baking afternoons disappeared. It wasn't quite the same any more."

"I've tried some of those vegan cakes, they're quite..." Peter started, but Maj didn't seem to hear him.

"The animal rights groups," she continued, "they are much more aggressive in England, aren't they?"

"Some of them are."

Peter told Maj the little he knew about the Eden group. He described his visit to the Roots and Shoots café. They talked a bit more about animal rights, then about Tina's other interests. Maj told Peter about her bachelor's degree in social sciences. He had heard most of this already but did not interrupt. Maj seemed happy to keep on talking. She told him more about Tina's school years. Apparently, her interest in the identity of her father was not a new thing. Maj admitted to making up a story about an English-speaking visitor, long gone. She had told Tina that she didn't even know his name, but that he had been terribly charming. The story got a life of it's own. Together, they fantasized about who he was and where he was now. A prince, a spy, a great doctor, hidden in anonymity. For a while, third or fourth grade, Tina would only speak English.

"If she didn't know the right words, she just made some up. It was quite amazing. And funny, sometimes." Maj smiled at the memory. "Christina started speaking in full sentences way before she turned two. She's fantastic with languages." A proud smile emerged, but was just as quickly tempered. "Luckily, she lost the absent-father fixation after that. It remained a childhood story, a family myth, not exactly believed, but not directly questioned, either."

"Until recently." He said. She nodded but pursed her lips and didn't volunteer more. He found another topic.

Peter left after a couple of hours with his head full of stories and bits of new information. But he had also become much less sure how well he actually knew Tina. Not all things fit. As he walked back along the early evening streets, he remembered the many things he had forgotten to ask Maj. He still did not know what their big argument had been about and whether it might have anything to do with why Tina had stopped seeing him as well. And what exactly had Maj told Tina about him? This question made him realize that Maj had asked him nothing about himself. She knew the professional side, but he had not added anything personal. Either she did not care, or Anna had already filled her in. Well, he had to hope that this was a start, not an end.

His phone rang. It was not Maj, but Anna. She wanted to invite him over to look at some pictures she had found. He accepted happily for the next day. He had one additional assignation in Copenhagen and he preferred not to have too much time for it.

———

The S-train heading north seemed full for a Sunday afternoon. It was yet another pretty Indian summer day. A conspiracy, Peter thought, to keep out dark thoughts. He smiled. He looked at the two pictures Anna had let him keep. One was Tina as a baby, old enough to sit but maybe not to walk. She was wearing a bright yellow jumpsuit with green dots and seemed to be biting down on a soft toy. Her expression could almost be recognized: skeptical but curious, determined to understand the fascinating object in her hands. Maybe the toy just tasted funny. The second picture was Tina at about eight, blond pig tails and freckles, sitting in the rabbit pen that Svend had built and smiling at the photographer. She had her pet rabbit, Pelle, in her lap. He could be identified by the large asymmetric patch of black on his otherwise white face, Svend had told him. "A pirate" Svend had suggested to Tina "should have a fierce name". She had insisted on the less combative name. Svend had shown Peter the pen this morning. It was still in working order but now filled with chickens. For eggs, Svend had explained. They couldn't bear to get more rabbits after Pelle died. He had also seen the vegetable patch Svend had been working on the day before. Gleaming red tomatoes, beans, zucchini and red beets recently dug up: ten square meters very well tended. Peter hoped he had expressed sufficient admiration for the produce. In truth, he had been distracted. The many pictures from Tina's childhood had overwhelmed him. Luckily, Anna had kept on talking as she passed him one after the other. All he had to do was mumble an adequate reply. He was content with keeping only two. The whole set, the full evidence of a life he had not witnessed, would have been wrong. Not that it was offered. They parted on good terms and agreed to keep in touch. It was understood that meant let each other know as soon as someone heard from Tina. Email addresses were added to the phone numbers. He had not spoken with Maj again.

—

"Peter. How good to see you." The older man moved forward as if for a hug, but stopped at a firm grasp on both of Peter's shoulders.

"Erik." Peter said, with less conviction. But he did smile.

They went inside and settled next to a bay window overlooking the garden. Peter sat in a small sofa, Erik in a low chair across from him. The house was still the same, but redecorated at least once since last he had been there.

"It's been a long time." Erik said. "Too long."

"I'm sorry." Peter said. "You know how it is—busy—too many things to do at work. I suppose I should have. . .."

"No, it's fine. How was the conference?"

"Fine." He had momentarily forgotten the white lie he had told to explain his sudden visit to Copenhagen. "It was fine. Going back tonight."

They looked at each other. Both smiled, but cautiously. The words didn't come. Erik cleared his throat. Then a door opened and a woman came through with coffee and cups on a tray. Peter recognized her and yet didn't. It had been over twenty years since he had last seen Laura. And even then it had been brief, and willfully unacknowledged from his side. It was at the funeral and she shouldn't have been there. He still hadn't forgiven his father for that, he realized. She was a striking woman, but very much changed from the dazzling young student who had lured away his father when he was ten. Her dark hair was streaked with gray, but expertly cut. The clothes fit her still-slim body very neatly, slacks and a blouse, conservatively dull but probably very expensive. The jewelry was discreet. Her face was perfectly made up, as different from Anna's weathered wrinkles as it seemed possible. Laura sat down, also in a chair, and poured coffee for all three of them. The many years of largely unvoiced accusations sat heavy in the air. His father looked old now, Peter realized. He looked his age, actually. Peter forced himself to re-inhabit his adulthood.

"So, how is Mikkel?" He said, with what he hoped passed as genuine interest. He looked at his father and then at Laura. "And Sofie?" He added, leaning forward and picking up his cup. The grandchildren's names, his half-nieces and half-nephews, he had forgotten, but he felt confident they would be provided along the way. They were. His father looked visibly relieved and started answering eagerly. Laura kept a more neutral stance, but added comments and details along the way. Peter got plenty of information but retained very little of it. Laura asked about Jessie at one point and he gave a few bland, non-informative answers. They had never met, so it was not important. His father initially seemed to have forgotten who Jessie was, but quickly recovered and almost managed to hide it. While they were talking, Peter observed him carefully, the short, gray hair, the sagging cheeks, the generous bulging at the waist and the slightly hesitant movements. He felt moments of sad pity alternate with moments of dislike, almost disgust. Both of these feelings were inappropriate, he thought, but nothing better was available to banish them.

"I'm afraid I have to catch a train back in half an hour." Peter said, looking at his phone. "I wouldn't want to miss my plane." He smiled, knowing it to be forced.

"Well, we can't argue with that." Laura said, relaxing her tensed shoulders. Relief, Peter assumed.

"Would you like me to drive you?" Erik said. "To the airport, I mean."

"No need. It's so convenient now, with both train and metro."

They got up. Laura started clearing the table and Erik followed Peter to the door. They still had a few minutes.

"So how is retirement treating you?" Peter said, in a voice noticeably warmer than before. "Are you finding enough things to do?"

"Well, you know, I still keep a hand in, where I can. I'm on a few boards and advisory committees. These new companies like to have a man from big Pharma around, even an old man like me. I suppose it helps that I'm no longer working for the competition. I can dole out the odd bit of wisdom. Or simple common sense." He smiled. "And the grandkids, of course. There's always something going on with them. Noah, you know, has... Well, anyway... Laura does most of that."

"Good. I'm glad you're keeping busy." Peter said. Erik opened the front door. This time, they simply shook hands.

"And your bag?" Erik asked as Peter was leaving.

"Yes, of course, I still have to pick it up at the hotel. So I'd better run now. I'll email you."

"Yes, yes." Erik said, but shook his head. "You are always welcome, you know." He looked about to say something more serious, but just paused and added. "Give our greetings to your wife. We hope to meet her some day."

Peter drew a deep breath and started walking toward the station. He checked his phone. There was still nothing from Maj. He was not sure why he expected there to be. He thought of calling Jessie. He decided not to. He needed time to digest all this.

———

Once he was on the train going downtown, he pulled out the two pictures again. He looked at them, but his mind drifted to where it had wanted to go since yesterday. Now he would let it. But he would not go to the cemetery. It was a cold and lonely place, a place of death, not life. He didn't need a headstone to remind him who Birthe Dahl, beloved mother, was.

He went one station past Nørreport and walked up the familiar steps. He turned right, then left, then right again. The traffic was modest, but there was plenty of street life, a lot more life, in fact, than when they had lived here. Outdoor cafés were doing bustling business; several little shops were Sunday-open. The yuppies outnumbered the druggies by a lot now. He walked further, turned two corners. The buildings were the same, though. He stopped at the corner and looked up at their windows. The wall behind him still had some afternoon sun. He found a good spot, leaned against it and looked up once again. He thought of the old-fashioned apartment, quite large and spacious for the two of them. He walked through it in his mind, first his own room in one of its many incarnations, then the narrow hallway and the two big living rooms connected by always-open double doors. He walked along the shiny wooden floors, touching the books in the bookcases lining the wall. He walked behind the big pale sofa and the well-worn leather chair, her favorite chair. He went into the kitchen, a large kitchen for such an old apartment, with a central table always crowded with life: fruit and flowers from the market, his schoolbooks and her science journals. She was sitting right there, close to the window, and looked up with that special smile, the smile just for him, as he came in. She was younger than his age now, much younger. It was Sunday and they had all the time in the world. Where would they go? The parks, the Zoo, one of the museums, the canal and parliament? Perhaps even Tivoli? Whatever he wanted, she would say. He would look at her and guess what she wanted. A bigger smile told him when he got it right. On their long walks, they would look at everything carefully. She would explain when he asked. She would listen patiently to his outrageous alternative explanations, made up on the spot, and she would laugh, calling him little professor, my young gentleman and sweetheart.

"I miss you so." He said gently and quietly to the windows above.

On the hour-long walk back to his hotel he continued speaking softly, under his breath. "You have a granddaughter," he started. He told her everything about Tina. He told her that she would like her. He was sure she would.

* * *

The lab was quiet. Ilana was unhappy and not trying to hide it. Peter felt her discontent being actively beamed in his direction. Whenever she wasn't in Carol's lab or down in the facility, she sat at her desk, elbows up, staring at the screen. Occasionally she'd be doing something on the computer but she hid it whenever he got close. He knew that he should talk to her, and he also knew that he was putting it off, for one good and one bad reason. The bad reason was painfully simple: He knew what was wrong but he didn't want to have it confirmed. The good reason was pedagogical: It shouldn't always be him chasing her. She should come to him with her results. She should want to discuss them—like Mihai, always eager to talk. Being secretive about a result as he was now, that was unusual. And he was obviously doing it for the drama of the unveiling, not to hide behind. Mihai was in the flyroom now. Perhaps he could. . .

His mobile rang. A Danish number.

"Hi—I'm sorry for disturbing you—I guess you are at work?" It was Maj. "Can you call me back? We are not suppose to use these phones for. . ."

"Sure, just a minute." Peter closed the office door, picked up the desk phone and punched in the number displayed on his mobile. He had not gotten around to switching mobile providers. It was criminal what his present one charged for international calls.

"So, have you heard anything?" He asked, after they had gotten past the repeated apologies.

"No. I was hoping you had."

"Not yet. I'll call if I do, as promised."

"Good. Fine. . . I just. . ." He waited. Silence.

"Is there something I should know? About Tina?" He guessed.

"I was just sitting here, talking to a patient, and it reminded me. . ."

"Yes?"

"What was Christina's mood like, when you last saw her?"

"She was a bit down, it seemed. She didn't say very much so I never found out what was wrong."

"And the weeks before that? Was she happy, active, a bit restless?"

"Yes. Once she had shed her initial reserve, she was quite entertaining to be with. She talked a lot, told stories and so on. Which was why I noticed the change."

"I should tell you something about myself, something that might also be relevant for Christina. You deserve to know, I think."

He waited.

"I was diagnosed as bipolar in my early twenties." She said, calmly, using what was probably her professional voice. "I don't want to go into details. It's not important for this. I was put on lithium and it really helped me. I've been stable ever since." She paused. He let her. "It's just, with Christina, I worry she has the same tendency." Another pause. "It's genetic, you know."

"A predisposition for it, yes, but no simple inheritance, as far as I understand."

"That's right. But she's shown signs of it before. And the mood pattern you describe. . .."

"Yes. . ."

"It sounds like she might be in a depressive phase now, after the slight mania you seem to have experienced."

"Shit." He said, lacking a better response. "That's not good—on her own in a foreign country. What would she do?"

"I don't know. That's why I'm worried." She must have been worried for the past two years, he thought, but decided not to pursue it, not now. "At home she'd stay in her room or go to my parents'. She'd hide under a quilt and eat only when something was brought to her. Christina needs looking after."

"She's not—you don't think—"

"What?"

"She wouldn't harm herself, would she?"

"No, no. It's never been that bad. She has her black periods, but not pitch-black. It's never gotten that extreme, I've always. . ."

"Always what?"

There was a pause at the other end. "Always—kept an eye out. Christina is not a danger to herself. She is just vulnerable."

They were both silent for a while.

"What do you want me to do?" He finally said, having swallowed all the other things he would have liked to say.

"Anything you can to find her. Bring her home if she is willing." She paused. "If not, just be there for her."

"OK." He thought of the Eden group, the café, his meager leads.

"Should I come over?" She asked.

"No." He said quickly. "I'll try to find her." It had been almost two weeks since his trip to Copenhagen, a month since he last saw her. "I'll be in touch." He rang off and sat looking at the silent phone for a while.

Chapter 8

"Hi Bear." She said.

"Hi Sweetie." His voice was soft, warm.

"You're there." She added, unnecessarily.

Jessie could see herself smile in the small insert on the screen. There was an empty wall behind her and a slice of the giant window full of green.

"Turn you camera on." She said. "So I can see you."

Peter looked slightly disheveled, his hair out of place, his eyes unfocused. Behind him were shelves of labeled folders and old journals, the edge of a door. He was in his office.

"Hi there." He said, and waved. His smile seemed genuine but was brief.

"Hi there." She held her smile, cocked her head. "I've been trying for a few nights, around eight or nine your time, but didn't... Well, anyway, I thought I'd catch you now instead." She paused, waiting. "Can you talk or is this a bad time?"

"It's fine." He got up and closed the office door. "I was just... I guess I should have told you. I've been out in the evenings, out looking for Tina."

"I thought..." She said, drawing it out, puzzlement with a touch of skepticism. "I thought you didn't know where..."

"I don't, really, but I got some pointers from Emily. You remember, the girl I told you about, from the Eden group?"

"The roommate. So she's trying to help?"

"She is."

She waited for more but it didn't come. Finally she asked. "Why now? It's been over a month, hasn't it? Maybe she has just..."

"Maj, Tina's mother, called me a few days ago. She's worried about her and asked me to try a bit harder."

"Now she's worried? After two years? I find that a little..."

"Well," he interrupted, "it's complicated." Again, he didn't expand. "So how are things at Oak Hill?" He said instead. She didn't answer immediately. He continued. "Did you finish the first round of interviews?"

She decided to let it be. "Things are great." She said. "It's super interesting." Her voice was deliberately cheerful, almost impersonal. He didn't seem to notice. "And, yes," she continued, "we just finished the first round. Fantastic talks. The mood at the interviews was so upbeat, so excited. I told you about that, didn't I?" She paused briefly. He didn't reply. "We've already decided on three of them: two use photo-activation and quite cool imaging stuff to study specific brain networks, one in mouse and one in fish. The last one is developing new electrophysiology probes."

"Three in one round? That's a lot at once, isn't it?"

"Not at all. We'll probably go for one or two more. They're all just starting out, and we have to..." She shook her head and continued. "Anyway, it's not like the standards are being lowered. The applicant pool was simply stellar. Didn't I tell...?" She heard herself faltering again. It irritated her. She forced herself to push ahead. "We plan to have another round of interviews next year. After the other things have gotten started." She took a deep breath. "We have the first Spark meeting next week."

"Spark?"

"My publishing thing. We decided to call it Spark. Catchy, no?" She smiled and tried for a slightly ironic tone.

"That was fast."

"It'll be sort of a trial run. Some of the presenters already have an affiliation with Oak Hill, or with Tony. They're all very curious about Spark. Next meeting will be late October. It's all pretty exiting, I have to admit." She flashed a quick smile. "It'll be two full days, twelve presentations. I'll be in charge of..." She stopped and continued more slowly. "Well, we'll see how it goes." He didn't respond. She looked at him on the screen. He was looking at the table, or at his hands. She decided to play it safe. "So how are things in the lab? Did Ilana get more numbskull results? And the new mutant from Mihai's screen—has he told you what it is yet?"

"It takes time, you know, to do actual experiments. Real progress is slow." The bitterness was unexpected.

"Right, sure, I know that." She could feel herself recoil even as she said this. "I was just asking because..."

"I know, I know. And I'm sorry. It's just... Everything, right now." He took a deep breath and tried looking directly at her. The screen interface made it awkward. "Now I've started worrying about the review as well. I've had no major papers in the last couple of years. The mouse project might be a bust, which means the whole drug angle is out. I have to tell Gerald. I know he was hoping for something there."

"Well, he can't be expecting that. Top papers every year. Or drug leads. No one knows when..."

"Of course. I was just hoping for little something special—something to help justify my existence, my choice of model organisms, basic science in general— preferably all at once. Well..."

"Have you talked to Hans about it, or to Carol? About the review, I mean."

"No, not really, why?"

"Maybe they have the same worries. It might be. . ."

"It's different for them. Carol, she. . ."

"Right. Yes. Well. . ." She frowned. "How about doing something with Hans? A boys' night out. The pub you both like. That always used to cheer you up."

"But there's Alessandra and the baby. I don't want to get in the middle of that."

"I'm sure he's allowed a night out occasionally. I mean. . ."

She let the words stop.

After a moment he straightened up. "I need to. . ." he looked sideways. Then he shifted his gaze back to the screen. His expression seemed almost normal, but forced. "And how are you? Are you OK there? After work, I mean."

"I'm fine." It was an automatic response, but close enough to the truth, she thought. "I'm still in the guest facility. I'll be moving to my assigned housing on Friday. I hardly have time to think about it. Too busy."

She left a pause.

"I miss you." She said, making a sad face at the screen.

"I miss you too."

Another pause.

"Will you be home one evening, so I can Skype you there?"

"Sure. . ." A grimace. "I just want to try a few more places. She's got to be somewhere. I've been walking all around London. It's endless, you know?" He shook his head. "Good sleep medicine, if nothing else."

"I'll just look out for you being online. You can also Skype me. Any time, you know? I'm not exactly overrun with people who need to talk to me." She gestured into the bare space of the office.

"Sure. . ."

"Love you."

"Love you too."

* * *

Tony and Jessie exchanged wide-eyed glances. Jessie had just called a formal end to the meeting. No more comments would be transcribed. Her assistants, Sharon and Erica, had their fingers off the keyboards and she had stopped the recording. She looked around the room, at the slightly flushed faces. She thanked them for their contributions, the thought-provoking observations as well as the insightful questions and comments. She explained once again what would be transcribed from the meeting and how the observation figures and their legends, along with all associated comments, would be circulated—and what they should do with them.

"Remember, factual corrections only. Deletions if you must." A "Boo" was heard from the other end of the table, then good-natured laughter. "And corrections to your own words only, of course, not those of others." More laughter. "No additions after the fact. We'll get all this to you in two days. You'll have two days to respond."

"Yes, Ma'am." This came from Sandy. He was at least twenty years her senior.

"OK." She smiled at Sandy and at the others. "It's a wrap."

Fresh discussions immediately broke out amongst the twelve participants around the table. Bodies shifted, heads leaned together, sideways or across the table. Two or three got up from their seats and switched positions. Loose paper was scribbled and drawn upon; laptops were opened to images and numbers. Eager squabbling filled the room. After a few minutes, Tony got to his feet, tapping an empty glass with his pen. They fell silent.

"I'd like to thank you all again for coming to Oak Hill Research Center and, most of all, for being willing to participate in this groundbreaking experiment. This has been the first ever Spark meeting. But the first of many."

"Hear, hear."

"These two days have been a real eye-opener for me. As Jessie said, your contributions have exceeded all expectations. But then again, exceeded expectations is what you all do on a routine basis." Softer laughter, from all around. "No, seriously..." Tony's speech went on for a little while longer and even became somewhat emotional. Applause broke out when he stopped. He indicated Jessie and said a few additional, very flattering, words about her. She lifted an inch or two in her seat and nodded. More applause. Then the twelve went back to their squabbling. Tony whispered to Jessie.

"Wow."

"Wow yourself." She whispered back.

"My office? And leave them to it? We'll see them at dinner."

"Sure." She smiled.

Jessie gave a few instructions to the assistants. They packed up their computers, notes and the tape recorder. Tony and Jessie left the room, nodding at Susan, the only participant who noticed their departure. In the corridor outside, they exchanged looks again. "That was..." Tony started. She nodded. It was. He turned and walked toward the main building. For a few seconds, Jessie remained in the glass corridor, soaking up the amazing place and smiling to herself. Then she followed Tony.

As soon as she was inside, Tony closed the door to his office and opened a small fridge behind it. He lifted up what looked like a bottle of Champagne.

"I saved it for today." He said. "We deserve it. You deserve it."

"We deserve it. Absolutely." She saw glasses on a nearby shelf and picked them up while he dealt with the bottle. It gave a muffled pop as the cork slid out and released a wisp of discreet vapor. Tony poured for her and for himself. They toasted and drank slowly, pensively.

"It was clear already yesterday, wasn't it?"

"It was." She smiled.

"Sandy and a couple of the others congratulated me at dinner. They were very enthusiastic. But I didn't want to jinx it by celebrating too soon."

"It was even better today, I thought. Everyone was participating, all the time."

"Practically, yes. They upped their game from yesterday. Once they realized at what level many of the comments were..."

"Did you give specific instructions to some of them beforehand? I'm thinking of Junko and Mengyao. Your friend Sandy, as well."

"I might have impressed upon Junko and Mengyao how important it was to get good questions and comments flowing from the start. Sandy didn't need prompting—he was in his element." Tony grinned. "Good choice of who to put first, by the way. Did you know his stuff was going to be so cool? It's shocking, really. Who would have thought. . ."

"I suppose I did know. I looked through the contributions last week—and I've heard Dubov give talks before, so I knew the presentation would be engaging."

"Right, right."

"I also had a chance to talk to Junko and learn more about her work these past weeks. Amazing stuff. She's got a totally different angle from anyone else, it seems. She. . ." She paused, briefly.

"I know. Our luck, isn't it?"

"Yes." Jessie smiled. "Anyway, I figured Dubov's observation about how the placement of synapses seemed to shape the response had to tickle her. And it did. Her questions were spot on."

"As were his answers."

"I really liked her suggestion about interference."

"Yes, thoughtful, wasn't it? And that comment from someone about resonance. Totally left-field. It got a bit wild."

"We couldn't have scripted it better."

"I know. And we didn't. It's the Spark. It really works."

They smiled at each other.

"And that was just the first observation." She said, happily. "Junko's own. . ."

"And Susan's and Mengyao's and what's-his-face from UCSF?"

"Alan Berger. Up-and-coming."

"I'd say. I was totally surprised by his evidence for these two different types of circuits."

"As were the others, obviously. Sandy spotted a caveat or two. That discussion was fantastic. I saw Berger taking notes feverishly."

"I guess he couldn't wait until the comments are published online. When is that, by the way? In two weeks?"

"That was the plan. But it'll be ready before that, I think. We have to make a big splash, so people will notice it and get excited about it, the science as well as the approach. I am dead curious about how people will react. "

"We'd better release it on the day we planned, then, not before. I'll confirm with the PR people. I'll get quotable reactions from the people we talked about last week, if you can have the final transcript ready a couple of days beforehand."

"I'll make sure you have it end of next week. I'll be giving it my full attention, as will Sharon and Erica. We'll follow up with the participants actively, calling if needed. But I think everyone will be responsive. They're excited about it as well."

"The room was thick with it, wasn't it?"

"It was." She looked straight at Tony, her forefinger tapping her glass for emphasis. "It's not unlikely that we'll be inundated with requests to participate in upcoming meetings when we go live."

"Which may create some tension. But the buzz will be worth it."

"Yes. We'll just have to deal with it." She said. "Luckily, I'm pretty good at that: managing people's expectations and feelings of entitlement."

"I know." He said. "That's why you are the perfect person to spearhead this. Apart from it being your idea, of course." She lifted her eyebrows. "And let's not forget your impressive knowledge of all things neuroscience."

"Yeah, yeah. . . But you believed in it and you got our first participants to do the same. They made it happen."

"They did." He smiled. "So how are things shaping up for the next one?"

She put her glass down, pulled out her phone and found the note. The program was almost full. He added another suggestion. She didn't recognize the name. He put his glass down and they moved to his computer so he could show her. They spoke rapidly, listened attentively, surfing easily on the success of the past two days.

At one point, Jessie noticed the family photo. She had never been on this side of his desk before, so it was new to her. The picture showed Tony, his wife and their two children. All but Tony were seated, Tony's wife in the middle, a child on either side. A skinny boy of about twelve looked uncomfortable, or restless, leaning away from the rest of the small group like he might take off. On the right was a girl, smiling and looking straight at the camera. Her age was hard to determine, but Jessie guessed she was younger than the boy. She was clearly a Down's child.

"Angie is such a happy child." He could tell what had caught her eye. "Always a ray of sunshine." He picked up the photograph. "It takes a lot of work, though, keeping her at home. And Ryan, he. . . Well. . . It's not always easy. Hester is our rock." He put the photo back down, but still facing them. "Now you understand why I can't travel much. I'll be relying on you."

Jessie nodded. She recognized the name, Hester. In the awkwardly staged photo, she looked tense: Pursed lips, severe hair, straight gaze and a hand going sideways to the girl, who didn't seem to need calming. Tony had a hand on his wife's shoulder. Jessie looked away.

The large window to the grounds and the forest beyond held a soft, fading light. It was still too early for real autumnal colors. But it would be beautiful, she guessed. She could see two figures moving across the grass toward the main building, deep in conversation. Tony noticed as well.

"Time for dinner, it seems." He said. She nodded. They went out to join the cheerful voices in the main foyer.

* * *

She knew he had a good reason for not being at home. A completely natural reason, one might call it, others might call it. But she still wanted him to be there, in their house, missing her, waiting for her call. Of course it was petty. She knew that. She chided herself. She didn't desperately need his attention. She wasn't twenty-four and lost in a foreign city. She was forty-six, an adult, with lots to do. Checking the many modifications and corrections to the transcript, for example. It had to be done carefully, with proper understanding of the subject. She would finish the work herself. She had the rest of the day. Twenty rings, still no answer. She clicked the red hang-up button.

Her view was almost as spectacular as Tony's. Today, an unusually strong wind gripped the trees. She watched them for a while, until she felt better. Looking back on her screen, she noticed Robert was online. On a whim, she called him.

"J.J., my favorite outlaw."

"Silly. No one calls me that any more."

"They should." He smiled, mildly pixilated. His office was packed with books, as it should be. Hers was still mostly empty. "OK, favorite sister then."

"And that's not allowed. Having favorites."

"For Mom and Dad, maybe not. For me, sure it is."

She smiled, as intended.

"I was looking out my window and reminded of West Creek Lake."

"Go on, make me jealous. My promotion still has me in the basement."

"Your literary cave."

"I suppose. But yes, please do come. The trees have already started turning out there. It'll be extra colorful this year, they promise."

"They always say that. So who's going—and when?"

"Mom and Dad from the tenth I think. Marion and I will be there for the weekend of the fifteenth. We're flying to BWI and driving from there. Nick and Bea will be home alone for a whole weekend. Scary."

"Thank God for mobile phones."

"Still. . ."

"It sounds like they are sensible enough to handle it."

"They are. So will you come? At least for the weekend? Mom and Dad will be thrilled if you do. I know they will. Me too."

"Let me see." She clicked to her calendar but already knew that the weekends would be empty. It was two and a half weeks away. "Maybe." She answered. He pouted. "Probably." She added and smiled. He nodded eagerly. "OK, OK, I'll come." She finally said, shaking her head and smiling some more.

"Excellent. It's so good to have you back stateside, Sis." He paused. "It'll be fine, I promise. I also promise to take you for long walks when they drive you crazy."

"Which reminds me. Is Kathy coming?"

"I don't think so. Driving all that way by herself, with the little ones in the back? Especially the terror."

"But Mom and Dad might. . . ."

"Five in a car, including the terror? That won't happen. I think you are safe."

"You shouldn't call him that, you know." She smiled, anyway. "The terror."

"Even Marion calls him that." He said, as if that settled everything. "Besides, you haven't met him yet."

"I haven't. I was hoping to postpone it to adulthood."

"Wise choice."

They chatted for a while longer, mostly about her job, a bit about his. Then one of his students knocked and immediately entered his office with an "Oops, sorry." The kid looked like an overwhelmed freshman in need of support. She rang off.

Now the office felt even emptier than before. Chicago was no longer a continent away—but it might as well have been. She wanted to pop around for a coffee with her chatty brother. She wouldn't even have minded a needy freshman barging in on her right now, demanding attention. But what needed her attention was right here, on the computer. She looked at the calendar in the foreground. She still hadn't bought her ticket to London. They had talked about early October, after the online launch. But everything seemed so uncertain now. He was never home. She closed the calendar view and opened the annotated documents from the first Spark meeting. This was important, and urgent. She had made the right choice, saying yes to Tony. She knew that. This was a real contribution to science. It mattered. She just had to keep it together.

—

It was completely dark when she finally left the building. It had been for hours. She had liked living in "her" guest room: The convenience, but also the feeling of impermanence. Now she had proper dwellings, a partially furnished apartment in the Oak Hill housing complex. It was just a mile away. The paths were well lit, for all seasons and all work-habits. But she turned toward the car park and the rental car awaiting her. Leased, technically. The simple act of driving again gave her such satisfaction: moving swiftly, going exactly where she chose, when she chose. It had been a small luxury during her university years, paid for by regular visits home. The visits never became very frequent, or very relaxed, but she kept up her end of the bargain. It had been part of her for many years, the car and the freedom it gave her. In London, they could both walk to work. Or take the tube. Or a bus. The little terraced house in Islington had stretched their budget even before the area became fashion-able. With her Oak Hill salary, it was a different equation now. Did that bother Peter? She didn't think so. She had told him what they had offered, of course. He had been impressed. The waiting car responded with frisky chirps and flashes to her thumb on the remote. She didn't want to think about the Islington house, not now. She started the car, her headlights flooding the dark entrance to the annex. She accelerated fast, deftly and with pleasure, up the winding and perfectly smooth access road.

A couple of minutes later, she passed the apartment complex. She kept going. Her mood was strange and unsettled; she was aware of this, but she did not know what to do about it. Fifteen minutes later, she pulled into the near-empty parking lot of a Giant supermarket. It was not the closest, but what did that matter? It was open.

The huge, brightly lit space was almost deserted. Apart from the staff, one at a till, one or two stocking the shelves, she saw only two other late-night shoppers. They walked alone and pushed shallow carts like hers. The muzak seemed loud. She walked purposefully to the large, open area on the right and picked up some vegetables and a few pieces of fruit, without looking too carefully. She couldn't eat every meal at the canteen. She noticed the corn, Silver Queen, and counted out half a dozen. She had missed that for years. Next came the dairy counter, where she found milk, butter and cheese. She got eggs, as well. An omelet would do for tonight. She moved on. There was an in-store bakery counter, not manned at this hour but with plenty of items on display. She squeezed a loaf of bread slightly. It felt crisp, yet with give. Even slightly warm—was that possible? The darker bread next to it felt softer, heavier. She took both. She noticed a brioche-like bread. She picked it up and quickly put it back down. Too soft, way too soft. Next to it were giant cinnamon buns, in packets of four. The butter-crusted swirls, golden in color and with generous dabs of icing, were calling to her. She closed her eyes and focused on her breathing. Eyes open again, she snatched the packet of four and started toward the tills at the opposite end of the store. The aisle she chose was the baked goods aisle. Ten types of precut white bread, followed by ten types of studier stuff. Next came transparent boxes of yellow, shiny-looking cornbread. It had been so long. She took one. A few steps along and it was all muffins and cupcakes, endless varieties, each as sets of four or six or ten, displayed and protected in their clear plastic trays. She quickly picked up a tray of oat muffins, then one of chocolate chocolate-chip. She spotted a tray of mixed cupcakes with orange and red and white icing, a special bulk deal, and some custard-filled things, also a special deal. Her shallow cart was almost full. She pushed it feverishly to the checkout counter. The girl who rang it up made no comment except for the total due. Jessie guessed it was a girl from the voice. She did not look closely and she did not make eye contact. She managed a quick and probably too loud no to double-bagging.

It felt good to be out of the chilly store. She breathed in and out, slowly. Although it was still warm, the wind was talking about autumn coming. The air on her face felt good, real. The car gave its loyal chirps of welcome. She bundled the ridiculous bags into the trunk of the car, closed it firmly and opened the front door. With her legs dangling in the open door, the car blinking an occasional gentle reminder of its unlocked state, she sat for a while, looking at the wind chasing an orphan plastic bag around the deserted lot. She listened to the sensible inner voice. She picked up her phone and typed in a search. It was not too far away and it was open for donations at all hours. She started the car and opened both front windows halfway. It was just a nice evening's drive, with lots of air.

"I can take the bread and the dairy and the eggs. If it's fresh, of course." The middle-aged, tired-looking woman who had introduced herself as Abigail was going through the bags. She glanced briefly at Jessie.

"It's all fresh." Jessie said. "Straight from the Giant."

"OK, then." Abigail produced a thin smile that gave nothing away. "Also these," she picked up the bag with corn and the bags with fruits and other vegetables, "but not this" she pointed at the last couple of bags, containing the buns, the muffins and the cakes. The unwanted bags remained, chastised, between Abigail and Jessie. "Too much processed sugar is bad for them, you know." Abigail said and shook her head. Jessie thought that a bit of fancifully processed sugar might be a nice treat for a homeless person, but she didn't argue. She picked up the rejected bags, not caring to soften the loud rustling of plastic within plastic, thanked Abigail with minimal eye contact and returned to the car she had parked—illegally—within clear view of the shelter's front door. When she turned back toward the squat, ugly building, Abigail had already gone inside with the donations. The street was deserted. It was a sad, neglected block of what used to be a real town. Jessie turned the car around quickly and accelerated.

At the apartment complex, she parked close to the elevator, picked up the bags from the trunk and locked the car. Home late, but otherwise everything was completely normal. On the third floor, she went past the door to her apartment and straight to the end of the hall. The building was so new that the round metal cover and the handle were still shiny, free of the grime that was sure to accumulate later. She opened the garbage chute and pressed first one, then another, bulging plastic bag into its expectant mouth. She saw a tray of muffins open from the hard squeeze and felt a pang of regret. Then she gave the bag another push and closed the cover. She returned to her apartment, fished out the keys from her pocket and went inside quickly. She got some water from the sink, but didn't bother to turn on the overhead light. It was well past midnight and only a few hours before she would be heading off for her usual session at the gym.

* * *

"Hi. You're there."

"I'm here." He cleared his throat. "How are you?"

"I'm. . . Turn the camera on so I can you." He did. His face was hazy, lit only by the bluish light of the screen. She couldn't make out anything else. "You're sitting in the dark." She said. "You're home."

"Yes. It's pouring down out there. It's not a night to. . ."

"I'm glad. That you're home, I mean." She looked down and bit her lip, stopping the other words from coming out. She missed him, terribly. She wanted so badly to be there, to be home, right this moment, wrapped in his arms, safe and warm and loved. And the rest of the world could just. . .

"It is a bit quiet here." He said. She thought she saw a sad smile.

"How are you doing?" She said, very softly.

"I'm still looking." He straightened up a bit. "She has to be somewhere. It's just tonight. I can't. . ."

She breathed in and out. Slowly. She would not cry. She would not react. She got her voice under control. "We had our first. . ." She started, then stopped. "I spoke to Robert yesterday."

"That's nice. How is he?"

"Fine. He's fine." Another pause. "I was thinking I'd go see them all at West Creek Lake in a couple of weeks. Mom, Dad, Robert and Marion. Just for a weekend. I'll drive there so I can. . ."

"West Creek Lake. That's the family cottage, isn't it? From when you were a kid? I remember you talking about the place."

"Yes. That's it." She smiled. "Still standing, apparently. It'll be beautiful there. The fall colors are so. . ." She took a deep breath and hurried ahead. "Would you like to come over? We could go up there together. Or we could make a separate trip, just the two of us. Maybe right after the next Spark meeting? You could. . ."

"I don't think so, sweetie. Now is not such a good time for me to. . ." He didn't finish the sentence. He didn't need to. She hadn't expected him to say yes, she told herself. Of course not.

"Well—how about I come home for a bit, then?" She asked. "A long weekend or something."

"That would be nice. Very nice, indeed." He smiled, warmly, naturally, a smile drawing in the whole face. "This place needs a little cheering up." He gestured to the dark room behind him.

"We'll light some more lights."

"Absolutely. It'll cheer me up, too. I know you have a lot going on over there. But I miss you, you know." He cocked his head. "Very much."

She smiled, happily. Relief flooded her body. The calendar was already up on her screen.

"Next weekend? From Thursday the sixth?"

"Sure. That would be wonderful."

"Good. I'll book the ticket. Today."

"I'm looking forward to it." He brought his face closer to the screen. It looked strange from the warped perspective. She tried not to think about it. He continued. "We could go out somewhere. Would you like to?"

"No. Just home. Let's just be at home. Next Thursday, I mean, next Friday, the seventh. I'll be there early Friday morning. Very early."

"OK, I'll be here."

"I'll bring breakfast."

"I'll be waiting."

"Love you."

"Love you."

Chapter 9

He sat in darkness while the rain kept rushing down outside. His smile faded only slowly. Next Friday morning—just one week and a few hours away. He needed it. Jessie had sounded like she did too.

He knew he should think about dinner, but he didn't have the energy.

A muffled but familiar ringtone startled him. He remembered that his mobile phone was still in the pocket of his raincoat, which was hanging by the door. The chair made a screeching sound as it was pushed back too forcefully.

"Hello?"

"Peter Dahl?" It was a female voice, young, but local. Not Tina.

"Yes, that's me."

"It's Emily." She paused. "From the café? Roots and Shoots?"

"Yes, of course. I'm sorry. I didn't recognize your voice."

"I'd hoped you'd be at the café tonight." Another pause. "She was here."

"Tina? You saw her? At the café?"

"Yes, she. . ."

"Is she still there?" He started pulling the raincoat off the hook. It was still wet.

"No, not anymore. But she was."

"But why didn't you. . ."

"You've been here most nights at closing. So I thought. . ."

"Yes. Sorry. Where did she go?"

"I don't know. She said she was looking for a place to crash. Then she left while I was serving someone else." After a moment, she added, apologetically. "She didn't look so good."

"Oh, no. . . ."

Of course, he thought to himself, the one night I don't go.

"I told her to come back tomorrow. Tonight I couldn't. . ."

"When tomorrow?"

"I don't know. But I'll be here until eight, as usual."

"If she's there early, keep her there. Tell her you've found a place she can stay. Please." He added, with more warmth.

———

He sat in his usual corner. It featured an actual chair and a crate for his cup. The closest table was out of reach. From this spot he could see anyone who entered the café, although mostly from behind. He had switched to herbal tea, what they called infusions, some time ago, trying his way through the variants. He had forgotten what most of them tasted like, but the idea of trying a new one each time appealed to him. Emily occasionally chatted with him when she was working, so the other girls accepted him. He was also clean, quiet and tipped generously. They probably assumed he was a like-minded local. He almost looked the part. He would read the pamphlets or a brought-along book for an hour or two, until closing time. Wednesdays he avoided. Today was Friday. Emily was behind the makeshift counter, glancing frequently at the door. Today, he was not reading anything.

Finally, after an hour's wait, Emily signaled to him with some urgency. The door had opened and a thin, hunched-over figure entered, wearing a plastic poncho with the hood down. Long, wet hair was plastered down her back. She headed straight for Emily, without looking around. He got up and moved toward the counter, in a way that blocked her direct route to the front door. This precaution turned out to be completely unnecessary. Tina did not try to run. She did not even seem very surprised to see him. Or perhaps the lack of reaction was just part of her general numbness. She didn't look drunk or crazy, she just didn't seem to be taking in what was going on around her. Vacant eyes, washed-out skin, thin, pale lips. He realized that he would not have recognized her had he walked past her on the street. She looked so unlike the spirited young woman who had approached him on a sunny June day. His eyes left her face. What he could make out beneath the ugly poncho looked in need of warmth, fresh clothes and food. He decided to keep it simple.

"Emily told me that you need a place to stay."

"Uh-huh." She half-nodded.

"I have a guest room. You can stay there."

She looked at his face quickly. Then she looked down and spoke softly. "But won't your wife. . ."

He wasn't really surprised at this. Anyone in her state would be shy of strangers. "She's in the US for a while." He said. He didn't move, keeping an unchallenging distance. "So I could use some company." His smile was steady and his expression, he hoped, welcoming.

She straightened up somewhat and looked at him again. Her expression changed, became a bit tenser and a bit more confident. There was a sudden flash of actual pride or defiance. He thought he recognized it, that flash. But he couldn't place it. Almost immediately, her expression was docile again. Maybe he had just imagined it. She needed help. She needed him.

Emily hailed a taxi for them. London's iconic black cabs had been part of the scenery for so many years of him walking everywhere. It seemed odd to finally step into one, a bit like a scene in a movie. He thanked Emily and gave the driver his address. Tina said nothing on the way there. Peter didn't mind. He would be patient.

* * *

"I'm sorry but I've to go now." Ilana and Carol both looked at him, Ilana with an expression of irritation, Carol with surprise. "Can we pick up this discussion again tomorrow?" Peter continued. They agreed, but with a reluctance that he seemed not to notice. They all started to get up. Ilana closed her laptop with more force than necessary and started noisily, still frowning, to pick up her papers. Carol closed her small notebook but kept her eyes on Peter. He switched off his computer but did not return her gaze. They left Peter's office and he locked it. With a nod to each of them, he turned and walked away quickly. Through the glass partitions, they could see him moving toward the central staircase. He did not look back.

"What's with him?" Carol asked Ilana once they were alone.

"You see? It's like he doesn't care any more. My project is going down the drain and he won't even try to rescue it. There must be other ways we. . ."

"That's not what I meant. I agree with him about Numbskull. We should respect the negative results. Accepting that your hypothesis is wrong can be hard, but it is important to acknowledge when to stop."

"Well, he's not the one who. . ." Ilana looked as if she had been wrongly accused of something, or been cheated.

"But he does seem very absentminded. Do you know if there is anything going on? Any problems in the lab?"

"I think his wife left him." Ilana said, with a shrug.

"Jessie left him? No way." This came out more forcefully than Carol had intended. "Says who?" She added.

There was a pause.

"Well." Ilana said, holding her laptop closer and looking uneasy. "A friend of mine works at the journal. She says Jessie Aitkin has recently quit as editor-in-chief. She has moved back to the US and taken a job there."

"Really?" Carol said, skeptically. "First I've heard of it."

"Plus he's here crazy hours. He leaves at five or six, but then he comes back later and he stays for hours—practically all night." She caught Carol's look. "I was doing some measurements every twelve hours, so I had to be here."

"Well. . ." Carol furrowed her brows, pensively. A moment later, she was brisk and businesslike again. "OK, so, we'll continue the discussion with Peter as soon as possible. I'll check with him to find a time. In the meantime, Huifen will show you the mutant strain I was talking about. I think she has the colony well expanded now, and the appropriate control strain as well. You can test them like you did the Numbskull mutant. You might as well get started."

"But. . . Shouldn't we ask Peter first?"

"He'll agree. Don't worry about that. I'll talk to Huifen, so she'll be expecting you."

Carol gave Ilana a friendly but firm nod. Then she too hurried down the central corridor. She did not take the stairs for her office one floor below, but continued all the way to the other end of the corridor. She knocked on an open door to announce her presence. Normally anyone would notice Carol approaching. She wasn't exactly a big woman, but she wasn't afraid of taking up space, physically, vocally. Hans, however, was so engrossed in something on his screen that the knock was necessary. He turned around and saw her. Before he could say anything she was inside and had closed the office door firmly behind her.

"Hans, I need your help."

"Sure, Carol. What's up?"

"It's about Peter."

"Peter?"

"Yes. You're his closest friend here, aren't you?"

"I'd like to think so, yes."

"So what's going on with him? Is it true that Jessie has left him?"

"What? No, of course not. I mean. . . I don't think so. I'd know. . . well, I'm pretty sure I'd know. We were at their place a couple of months ago. They seemed fine. . . normal." He paused. "But I admit I haven't talked to him much since then." He sighed. "I suppose I've been too wrapped up in my own stuff." Another pause. "Why are you asking?"

"Two reasons, really. Peter's Spanish postdoc, Ilana, just told me she heard that Jessie had left her job. Not only that, she may have moved to the US."

"That I did not know. Is she sure? Are you?"

"Let's look it up. Check the journal's website." She pointed at his screen.

He swirled the chair around and did as requested. They both looked at the screen. The journal had a new editor-in-chief, one of the long-serving senior editors. No sign of Jessie.

"What the Hell. . ." Hans started. "I had no idea. And where is she now?"

"I don't know. In the US somewhere."

He Googled her. A picture and a blurb about her appointment as scientific co-director at Oak Hill Research Center came up. The text also described the Spark publication project but neither of them read that far.

"Oak Hill." Carol said. "I've heard about it. Brand new. It's supposed to be quite something—best of everything." She paused and looked at Hans with narrowing eyes. "So. Has she left him—or just left first? Is Peter going to Oak Hill?"

"Not that I've heard of."

"It's either that or they've split up. Which is it?" She seemed to notice her tone. "Sorry, Hans. I shouldn't take it out on you. It's just. . . so sudden. But could you find out? Which it is, I mean?"

"Yeah, sure. I'd better talk to him." He looked at his hands on the keyboard but did not type anything. "I had no idea." After a moment, he looked back up at Carol. "So what was the other reason? You said there were two."

"Well, he seems to have been off his game for a little while, distracted. Today it was really obvious. We were discussing a collaborative project. Ilana has been testing the mouse Numbskull mutant for memory defects like the ones Peter found in the fly. She's been doing the mouse work in my lab."

"Right. And?"

"And—there's no phenotype."

"Really? That's too bad. He's been working on that pathway for ages."

"It's disappointing, of course. But there are lots of explanations worth looking at. It could be a false negative, due to redundancy, for example. Or it could really work differently in mice."

"Peter knows all that. He never oversells..."

"I know, I know. He was very calm and sensible about it. He's had some time to get used to the idea, as Ilana's negative results have accumulated. The weird thing was what came after: I told him about another mutant that we've been looking at. It's also a kinase, like Numbskull."

"Those names... And Dummkopf is the protease that acts afterwards." He shook his head, smiling. "I don't remember the name of the channel, though. We did some modeling of the system with Peter. We found that..."

"Yes, I reread the paper before our discussion—just to be sure I wasn't getting ahead of myself. Beautiful stuff, by the way. So, there's a conformational change of the channel, clustering, a set of non-linear phosphorylation events and finally the proteolytic cleavage, right? And the cleavage product gives the long term transcriptional effects."

"Right. Our modeling showed some really interesting properties of the system. For some parameters, it's very robust, for others.... It seems to be a molecular ratchet mechanism to build up molecular memory, but one that is sensitive to timing. It's quite cool." He was grinning.

"That's what I thought. The channel is very well conserved in the mouse and there's a whole family of proteases similar to Dummkopf expressed in the brain. Mouse Numbskull seems mainly to be expressed in the immune system. But—" she brandished her forefinger for effect. "—this other kinase we found might be doing Numbskull's job. It could be the same mechanism and logic you two worked out, just a slight shift in the components."

"That's great!" Hans smiled generously. "How did you find it?"

"Luck, really." She shook her head with an expression of disbelief. "It was in the kinase mutant collection. We've been testing a bunch of them for phenotypes." She smiled, broadly. "Anyway, I was sure that Peter would be thrilled. I was even prepared for a hug." Hans wasn't sure he'd dare to hug Carol, no matter how exciting the science. Her expression, however, had already shifted back to worry. "Peter barely responded to what I said. It was like it didn't even register."

"Really?" Hans looked skeptical.

"Exactly. I'm telling you, he's off in zombie-land somewhere. He just said he had to go. And then he left without another word."

"But—that's crazy. He should be all over it."

"So, Hans, please." She looked adamant in her intent, almost fierce. "Find out what's going on with him. Something is, for sure."

———

"May I come in?" Hans was at the front door. He smiled cautiously and continued. "You left early and you weren't answering your phone."

"Yes, of course." Peter shook his head as if trying to wake up. He was wearing an apron. "Come in, Hans. Come in." He started going downstairs. "Do you know anything about vegan food? I'm trying to cook these lentil things but they still seem hard as rocks."

"Vegan has always seemed a bit extreme to me." Hans said, starting after him. "Vegetarian food is difficult enough, at least for someone like me. Are you?" He stopped on the fourth or fifth step, surveying the downstairs space without really meaning to. Through the glass door he saw a person sitting in the wicker chair outside, in the sole wedge of sunshine. Even wrapped in a blanket, he could see it was a young woman, slender, with straight blond hair. Her face was turned sideways and she appeared not to have noticed his arrival.

"I spoke to Carol yesterday." Hans said, slowly and searchingly. "She was wondering. . . she was worried that. . ." He moved to the kitchen area and into Peter's line of sight. "Peter." He said. "Who is that, out there?" He pointed. "You haven't. . . have you?"

"That's Tina." Peter said. He took the pot off the heat and his apron off. "Hans, it's not what you think." Hans made a defensive gesture, professing innocence. "She's my daughter." Peter continued. "She's staying here for a bit."

"Your daughter?" Hans said, incredulous. "You have a daughter? And Jessie? Where's Jessie?"

"Yes, apparently, I have a grown-up daughter. It's complicated." He sighed. "Jessie knows about her." He looked at Hans for a moment, then turned to the fridge and opened it. "Let me get you a beer." Hans nodded. They moved to the dining room table and sat down. They each had a silent sip before Peter started talking.

"I suppose I should have told you, but. . ." Hans made a motion as if to protest. Peter waved it off. "But it's been a confusing time for me. So—where to start?" He took another sip. "Well, I'll start with Jessie. So. Jessie got an offer she couldn't refuse—" he grimaced, acknowledging the cliché, but needing it anyway, "—and off she went, to the US, faster than I-don't-know-what. It wasn't my idea."

"Oak Hill."

"So you know?"

"It's on their website. But why? Have you guys split up or what?"

"No, no. It's not like that. We're. . . OK, I think."

"So you're moving as well? Not the worst place to go, I suppose. When were you going to tell your old friends?" Hans tried for a smile, but didn't quite succeed.

"Well. . ." Peter shifted in his seat. "I haven't made any plans yet. That's why I haven't told anyone. To be honest, I don't know if I. . . It's not a great time for me to move." He stopped fidgeting and looked at Hans, steadily. "I suppose what I'm really hoping is that Jessie will do what she needs to do and, once that's out of her system, move back. Or something like that."

"So she gave up her brilliant job here, moved to another continent... And you think it's just a phase?" Hans sighed with intent. "That doesn't sound too good, Peter..."

"It's a very special opportunity for her. I couldn't deny her that, could I? With all that I have. And have had. So we'll see..."

They stayed like this for a while, sipping their beers, wordlessly. Peter glanced outside briefly when Tina moved her chair to follow the sun. Hans looked in the same direction.

"So," Hans said, "you have a daughter. You are full of surprises these days." He looked at the girl outside again, more carefully. "Wait, but isn't that...? Wasn't she one of the...?"

"Yes, one of the protestors from a couple of months back. They're called the Eden group." Hans started saying something, but Peter stopped him. "She's not really a member of that group, she's just... She's not even from here. She's from Copenhagen. I think she basically did all that to get my attention."

"Right, then. She's from Copenhagen. So you were..."

"Just don't tell Carol about this, OK? Not yet. Carol has no time for animal rights people."

"Well, for good reason. You know what they've done at other institutes, don't you?"

"Not Tina. She's just—she's drifting a bit, I think. That's partly why she's here. She's got no place to stay, right now." He paused. "I'm trying my best."

"So how come you never told me? It's not like I wouldn't understand."

"It's a strange story."

"Try me." Hans nodded, leaned forward.

Peter told him about Tina's initial approach, about the impromptu in-house paternity test, which Hans, to Peter's relief, found hilarious, and, after glancing over to check that the door to the patio was still closed, about his visit to Copenhagen, meeting Tina's grandparents and her mother.

"So you're saying you don't remember her—or making her pregnant—even after seeing her again?" Hans sounded doubtful. Not accusing, but doubtful.

"I've gone over it in my head, again and again. I have absolutely no recollection of it. Apparently, it was after a party." He furrowed his brow. "I never knew. She didn't tell me. If it wasn't for the DNA test..."

"Are you sure? I mean... I understand if..."

"Yes. Of course I'm sure." Peter said, obviously somewhat annoyed. "Or I wouldn't have... She didn't want me to know."

"Sorry, sorry." Hans held up his hands. "I didn't mean anything... I'm sure you would have been a great Dad."

Peter thought for a while before responding. "It's strange, you know, this sudden parenthood. Well, it's not parenthood, really, is it? Not like you've been a parent. It's somewhere in between that and being an anonymous sperm-donor. Closer to sperm-donor, I guess." He made a face. Hans shrugged. "Maybe it's just an illusion, this feeling, this connection. I don't know." Peter shook his head.

"It would be a shock to anyone. Especially if you don't have any real children. . ." Hans glanced at Peter quickly. Peter gestured to indicate no offence had been taken. "Children that you've been living with." Hans corrected. "Don't beat yourself up over it. Just do what feels right." He held up his empty beer bottle. Peter nodded and they moved to the kitchen to collect another round.

Tina must have noticed the movement inside, or else gotten cold. She uncurled from her position in the chair, got up slowly and moved over to open the door. Peter performed the introductions. Tina held tightly on to the blanket wrapped around her, even while she shook hands with Hans. She didn't quite look him in the eyes but she did manage a small "Hi" before retreating up the stairs.

"Sorry about that. She's a bit withdrawn right now. I was happy to see her sitting outside this afternoon when I came home. She's been sleeping for most of the four days that she's been here. But at least she eats what I cook. If it's vegan." He shrugged and gestured toward the abandoned pot. "It's damned difficult to make anything that's even remotely interesting." He made another face and they both laughed, mostly with relief. Then Peter's expression turned serious. He went across the room and looked up the stairway. When he came back he spoke in a lowered voice.

"Actually, her mother thinks she may have a real depression. There's bipolar disorder in the family. To be honest, I may be in over my head with this."

"Shouldn't she be at home, then?" Hans half-whispered. "I mean, I know you are trying to do the right thing here. But maybe her mother and her doctor, the people who know her best, should be taking care of her." Hans cocked his head, questioning. "Don't you think?"

"In principle, I agree. But Tina and her mother aren't on speaking terms right now."

"They're not? Why?"

"They haven't spoken for two years. Neither of them will tell me what it was about. Actually, Tina tried hard to keep her full name and the identity of her mother from me. She still doesn't know I've tracked Maj down and spoken with her. I'm afraid to tell her, in case she runs off again."

"Take my advice: Don't get in between girls—or young women—and their mothers. You should hear Elise and Marjorie." He opened his hands wide while he shook his head. The he stopped, and frowned. "Not that Elise is very nice to Alessandra, either, I have to admit. But Elise and I get along really well. She needs me. I can tell. I give her love and support, but without all the female fuss and fights. Maybe Tina needs you in the same way." Peter nodded. "So it's probably best not to tell her that you've been talking to her mother—at least for now. If you can manage that without lying outright, so much the better. You don't want her to catch you at that. Kids can be very unforgiving." Hans added, knowingly. Peter looked exasperated. "Just do your best. Be there. You'll be fine."

They sat for a while in silence. Then they started to talk about work, mostly about the upcoming review, while finishing their beers. Hans also told Peter to hurry up and talk to Carol again, but he wouldn't say why. This got Peter curious and he tried to get Hans to say more, which he politely, and teasingly, refused. Hans did, however, take Peter's reaction as a good sign, a sign of him getting back to normal. Shortly after, Hans was on his way home.

He hadn't told the whole truth, he realized. As soon as Hans had left, he sat down again and tried to think constructively about the omission. He had not told Jessie about Tina staying at their house. In a few days, Jessie would be coming home. She would not be expecting a houseguest.

* * *

"I'm down here." She called out. He pulled a laptop from his bag and carried it downstairs. She was in the kitchen. The aromas suggested serious cooking.

"Hi." He moved closer. "That smells fantastic."

"I thought I'd show you how nice vegan food can actually be." Tina smiled at him. She was completely transformed. He was too surprised to comment immediately. "So, who was that guy who came by yesterday?" She asked.

"Hans? He's a good friend—and a colleague."

"You talked for a long time."

"We've known each other for ages. He joined the institute shortly after I did."

"So he works on flies as well?"

"Hans? No, he's a computer guy. He builds models—on the computer." There were two pots on the stove, both covered, and something unidentifiable in the pan.

"A computer guy? I thought you all worked on biology."

"Hans does, as well, in his way. He makes mathematical models of biological events." He looked at her face, carefully, but saw nothing aside from what used to be her normal demeanor. "It helps us biologists determine how much we actually understand—and how much is just words." He continued. "If you can make a model of it, and the model behaves, you've probably got something. Sometimes the model does things we don't expect and we get new ideas that we can go and test in the lab. You'd like his approach. No animals." He reached for one of the pots. Lifting the lid, he burned his fingers, as he should have known he would. He let go quickly and it fell back with a loud clang. "And what's in here?"

"Careful." She said, smiling. "Spinach and chickpeas. It's a curry, or sort of. Some of the ingredients are missing." She looked mildly displeased, then cheery again. She pointed at another pot on the stove. "We'll have rice with it. And a red beet salad. It's already done—over there." She pointed to a bowl at the end of the counter. "These crispy snacks first, though." She flipped one of them on the pan, the oil sizzling.

"Well, it smells great. I'm looking forward to it." He looked at her working and he smiled. He just kept on smiling. It seemed to him that he had never in his life felt so relieved. It was not a serious depression after all. She must simply have been exhausted from living and sleeping—or not sleeping—on the streets. Asking about that time, and about why she ran out on him, could wait. "Thank you for doing this." He said instead. "Anything I can do to help?"

"Lay the table. And tell me you're ready to go vegan once you've tasted this. No, wait—tell me after I've had another chance to cook—with all the right spices. I need to go to a proper Asian shop. I know a good one."

"Sure. Of course. I had no idea that you'd. . . I'll go with you - tomorrow."

"OK." She flipped the golden patties on the pan one more time.

He had a moment of panic, not sure he knew how to keep the conversation flowing lightly. "But—but what about cheese?" He said, quickly. "I love my cheese." He thought of Wallace and Gromit but wasn't sure she'd know them. "What if I promise to buy only organic dairy products?" He was mock-pleading with her and she seemed to get it.

"That helps. A little. Happy cows." She flicked her ponytail. Putting her hair back up had also made a difference. She looked younger and healthier this way.

"I'm relieved to hear that. Anyway, you should teach me how to do some of this. It looks a bit complicated."

"Taste it first." She put the crisp patties onto some paper towel on a plate, but shooed away his fingers. "Not now. When they've cooled down a bit." She seemed to enjoy being in charge for once and he was happy to let her. He asked more questions about vegan food and about the spices she needed.

——

They were clearing away the dishes from dinner when the Skype call came in. Tina had let him work while she finished preparing their feast and he had left everything on.

"Hi there." Jessie's happy face beamed at him. "Camera on."

"Hi sweetie." He turned it on. "What's up? It's early for you to call."

"And you are home." She said, looking puzzled for a moment but then hurrying on. "I'm calling from work and I have to make it quick. I just wanted to tell you that I'll have to take the late flight tomorrow, so I won't be there in time for breakfast, but. . ."

"Yes, of course, Friday morning," he said, speaking clearly, a bit loudly, "you'll be back home. I'm looking so much forward to it."

She looked worried now. "What's going on, Peter? What are you doing?" Her eyes flicked back and forth. "And who is that in the background? Someone is there, Peter, in our kitchen. Who is it?"

"It's Tina. She made dinner." He paused. "Look, I'm sorry, Jessie, I should have told you. It all happened so fast."

"What happened to fast? What are you talking about?" She stared hard at the screen, becoming more agitated. "Peter? Talk to me."

"Well, I finally found her, a few days ago. She was sleeping rough, I think, but she made contact with Emily, and Emily called me. I was really worried about her, as you know. . ." He was speaking too fast. "So she's staying here for a little while, in the guest room. Just until. . ."

"Until what?" Her voice went up. "When were you going to tell me? Were you just going to let me come home—come home and find a stranger living in my house?"

"Keep your voice down." He hissed. "She's not a stranger, she's my daughter. And she had nowhere else to go."

"But you said you found her with. . ."

"Please. Not now. Let's talk about this when you get here." He said with an unexpected intensity, but the voice low. She pursed her lips but did not respond. "Be nice. Please?" He continued. "She's had a really hard time." He turned away from the screen and yelled toward the kitchen area.

"Tina—come say Hello to Jessie. You two should meet."

Jessie started saying something but stopped. Tina turned around with an expression of mild panic, then back to the stove. Peter picked up the laptop, held it to his chest and moved closer to Tina with an encouraging smile. She shook her head and looked dejected, all the lightness of earlier completely gone. He gestured to say "why?" She shrugged and looked down. Too much, too quickly, he guessed. Then she put down the pot she had been cleaning and moved toward the stairs. He turned the laptop over and talked to the screen.

"I'm sorry, Jessie. We can't do this right now." She started to say something but he cut her off. "I'll see you Friday, OK? I'll be here—I'll work from home. Have a good flight, sweetie. Yes? I love you."

It wasn't clear who cut the connection first.

—

He awoke to a phone ringing. His mobile. He had had the foresight to bring it to the bedroom. Computer and mobile. The lack of a landline had its advantages.

"Hello." His voice was fuzzy.

"Peter." Hers was not.

"Yes, I'm here."

"I recognize her. Tina. Your so-called daughter."

"She is my daughter. You know that. Don't be like this." He woke slowly. "Recognize her from where?"

"She was at my gym. Every morning. Almost every morning."

"Don't be ridiculous. You didn't even see her."

"I saw her almost every morning."

"I mean tonight. On Skype."

"I wasn't completely sure. So I checked for pictures on the web. It turns out all kinds of people have uploaded pictures from that protest in June. Tourists seem to find it interesting to take selfies in front of anything. Another animal rights group has posted some pictures as well. Not the Eden group themselves. They seem to be rather secretive. No names, no photos. Anyway, I'm sure it was her. She's the girl who was always there, at my gym."

He was having a hard time processing her words.

"I'll ask her about it in the morning."

"She was stalking me, Peter, for months. This was way before she made contact with you. It's really creepy."

"That's ridiculous. It was probably just some girl who looks like her. Or there's some other, perfectly reasonable explanation. I'll ask her in the morning."

"I want her out of there."

"What? No. Why?"

"Peter, I'm telling you. . ."

"Jessie, it's the middle of the night. You've gotten some weird idea into your head. I have a vulnerable girl on my hands and I'm trying my best here. She's finally opening up. . . We'll talk about it tomorrow."

"I need you to listen. You need. . ."

"Tomorrow. Goodnight." He clicked the end button and put the phone down. Despite everything, he fell right back into a deep sleep.

———

The next morning, he could almost convince himself the phone call had never happened.

It had just been a bad dream.

Except for the text message: "Trip to London cancelled. J."

Chapter 10

Her mobile phone was on the dash—not a great place for it. It slid back and forth as she pounded on the treadmill. She pushed the phone in under her towel and increased the speed from 8 to 8.5 mph. She imagined him lying awake in their bed, looking at the phone, wondering and worrying. It was almost an hour since she had called him. Seeing the large room full of treadmills and ellipticals had jogged her memory. There was a computer in the members' room. She had searched the Internet for images and she had found what she was looking for. As soon as she saw the girl's hair, posture and the side of her face, she was sure. And then she had to tell him.

Now she was trying to push everything away, by sheer effort. Going to the gym in the evening instead of the morning had been a brilliant idea. Habit had initially kept her in the early routine. But once she started on evenings, about a week ago, she wondered why she hadn't thought of it before. The day's events could be processed, the frustrations dealt with. Plus, she actually ran faster. Strange but true, she thought as she increased the speed to 9 mph. She was flying now. She concentrated on her breathing, on the rhythm of her feet, again on breathing, on her feet. Her earphones were off today, so she caught the snappy pop of the gym's sound system, mixed with noises from all around and the sound of her own feet, her own breathing.

Her left foot was hurting at the usual place, by the ankle. She pounded on, defying it. It would go away. This time, it didn't. It got worse, then sharply worse. She pressed stop and hobbled the last few steps.

"Shit." She exclaimed. "Shit, shit, shit."

She walked slowly away from the treadmill, each step punishing her.

"Hey!" A voice from behind her, a young woman's voice. Resonance of the London gym made her freeze, an eerie panic creeping in. She turned to see a plump girl in a tight-fitting black and pink outfit. She was waving a phone in the air. Jessie had forgotten it on the treadmill. She limped back and thanked the girl with as much grace as she could manage. The phone felt cold in her hand. Leaning against the protective arms of the treadmill, she sent Peter a brief message. It made her feel better for a moment. Then she tried walking again.

———

Painkillers and the sleeping pill meant for her overnight flight helped her into the next day. It was a busy day, as she knew it would be: The official online launch of the first Spark issue. Yesterday, the scientific press and the high end of the regular press had received previews, along with an explanatory write-up from her and Tony, and a few comments solicited from well-known supporters. This way, they could post their own commentaries on the day of launch—if they cared to. Several did. She saw the articles in the morning. They were positive about the initiative, some even enthusiastically so, and curious about the future of it. Calls for additional comments and from scientists interested in participating came in all day.

At lunchtime, she limped downstairs to meet up with Tony at their usual table in the canteen. Oak Hill was starting to fill up, but somehow everyone knew to leave this small table for Tony and Jessie and their guests. Today, however, they were not left alone. People came by to congratulate them, not just Junko and Mengyao, but also PhD students and administrators. Some stayed for a short chat. The excitement was affecting everyone. A one point, Tony clanged two glasses together for attention. He stood. Everyone in the canteen hushed.

"As you may or may not have noticed," laughter broke out, "we launched Spark publications today. This marks, we hope, a new era for science publishing and scientific discourse. Now, while Spark publications is supported one hundred percent by myself and by all of Oak Hill, I would like to remind you that the person who came up with this idea, and who is now spearheading the project, is our colleague, our scientific co-director, Jessie Aitkin." He turned and indicated Jessie. There was a prolonged applause before he continued. "So please join Jessie and me for a celebration of the launch in the annex foyer at five. See you there."

Jessie was touched, both by Tony's little speech and by the spontaneous goodwill all around them. He smiled at her and they finished their lunch. That there would be a celebration was a surprise to her, even if it shouldn't have been. Yesterday, Tony had told her that there would be an extraordinary faculty meeting at 5 pm today. That had obviously been a cover. She stayed in her office for the rest of the afternoon, answering calls, emails and friendly knocks on the open door. She heard nothing from Peter but almost succeeded in not thinking about it.

At five, she was surrounded by well-wishers.

At eight, she was sitting alone in her car outside the Giant supermarket.

The gym was out of the question. She had declined dinner with Tony and his family, as he probably knew she would. Sharon and Erica had suggested they go out somewhere after the Oak Hill reception. But she knew better. This was the unavoidable blank space after the limelight and the rush. She had to face it. Alone. Eventually, she had to go home. Alone. She turned off the ignition, locked the car and crossed the parking lot, limping.

The vegetables, fruit and dairy were ignored this time. There would be no visit to the stern Abigail at the shelter afterwards, either. Not this time. She went straight for the bakery, which, surprisingly, was still manned. She did not ask for anything, but picked up a loaf of bread, whole-grain and dense and still slightly warm, and a box of cinnamon buns, also warm. Next came the muffins, the cupcakes, the lemon sponge cake, carrot cake and the miniature brownies. She placed each item securely in the trolley. Her movements were calm, controlled and she did not linger. Finally, a quick survey of the trolley told her there was no point in getting more. On her way out, she passed the juice cooler and picked up half a gallon of orange juice. The girl at the checkout counter asked no questions.

"Ow, shit, shit, ow, ow." She exclaimed all the way to the car. Without the trolley for support, her foot was protesting again. The bags fit snuggly into the trunk of the car.

In front of her apartment, she put the bags down. She unlocked the door, stepped in and stood there for a moment, inside, looking at the bulging bags, still outside. She knew exactly what they contained. Once, her hand reached for the door handle, ready to close it and leave the bags stranded. But the quiet behind her was too relentless. She accepted the inevitable. She picked up the bags and brought them inside. At the low sofa table, the items were carefully unpacked and placed on the table, along with paper towels and a glass for the orange juice. The packaging was put away neatly in one of the empty grocery bags and placed next to the sofa. She turned on the TV to turn off the quiet.

Then she started. She tried the bread first, breaking off a large corner and biting into the center of it. It was dry and lacked in fullness; it would never do. She put the bread aside. Her next item was a cinnamon bun. She started from the outer swirl, but soon broke it in half to get at the moister center. It was satisfyingly sweet and chewy and sticky. She ate the whole thing. It was followed by an oat muffin, a banana muffin, some cake, a couple brownies and then back again to the cinnamon buns. She ate with determination and deliberation, not stuffing everything in at once, but also not stopping. The orange juice helped. To anyone observing, she would have looked like someone just getting on with the job at hand. She paid attention to the textures and the tastes, one item after the other. Eating the rich, sweet and dense goods did exactly what it was supposed to. It soothed her, satisfied her, filled her and blocked out everything else. The hollow and despairing feeling that yesterdays rupture had brought could not claim her now. From time to time, she glanced at the TV but without taking in the drama on the screen. She had what she needed right here, right in front of her.

She got through more than she had expected but less than she might have managed years back. And then it was time. She knew it was time. She sighed and got up slowly, careful not to touch her bloated stomach. Once in the bathroom, she put an elastic band around her hair, lifted the lid and got down on her knees.

That part was worse than she remembered it being. It was harder to make herself do it and harder to deal with the waves of shaking and hopelessness that came with every heaving. To hurry it along, she mixed a large glass of salt and warm water and drank the unpleasant mixture. It did what it was supposed to do.

Afterwards, her throat was sore and her fingers wet and a bit red. Her hand was still shaking. She got up gradually, and used her left arm to support herself at the sink while she washed her face and brushed her teeth twice.

In the living room, she slowly cleared away the remaining food items and put them in the bags with the packaging. This was taken to the garbage chute and discarded with little emotion. The hurt in her foot felt good and real. The juice was poured down the kitchen sink. The TV was turned off. She could still taste the sick and brushed her teeth once more. Then she removed most of her clothes and went to bed, sliding under the covers. She couldn't lie down immediately; it was too uncomfortable. So she sat up, propped up against two pillows, hoping sleep would claim her soon. She felt thoroughly awful, terribly stupid and very disappointed with herself—all completely as expected. She promised herself that this would not happen again. This was followed by a chastising acknowledgement that she would not keep this promise. She could not keep it. Not yet. Knowing this made her feel even more miserable. But she was not desperate. One advantage of no longer being young, she thought with resignation. She knew where this was going and what she would eventually do. She would get past it. She had done it before. But getting to that safe, firm, cool ground where she would have full control again required something she wasn't quite ready for. Not yet. She would have to live with this revolting urge, with the giving in to it and with the self-loathing afterwards. For a little while. Until she was ready.

Her train of thought ran to Rachel, her dentist, back in London. Rachel was a good dentist, a nice person, and, clearly, emotionally intelligent. She knew what to say and what to suggest for remedies, but also what not to say or ask about. They had an understanding. It was strange to think that Rachel was the only person who really knew about her, who knew everything. Her dentist. Well, no, her and the unfortunate roommates of her freshman year in college. But that was long, long ago and hopefully they had forgotten by now. She had put a stop to sharing living space, and, more importantly, bathroom space, as soon as she could. For the rest of her college years, she had avoided her former roommates. My God, she thought, those were awful years. Fighting with the overpowering, embarrassing urges, crying afterwards, not knowing if she'd ever be strong enough to win. Awful. Her studies didn't suffer, but... A few years later, she had succeeded. She was free of it. Well, almost.

"I thought so, too, Bear. But this was just too much. I needed... You understand, don't you?" She asked the scruffy teddy bear sitting next to her on the bed. It was the only sentimental item she had brought along from London. "You are not very tough, are you?" She caressed the bear's soft, old fur. "You just stay the way you are, alright? Soft and nice and cuddly. But I will have to be tough. Soon. It's the only way. I have to climb back up, all the way back up. This is so stupid. It's weak. It's disgusting. But I will get back up. You know that, don't you? It's just—we can't tell anyone—OK? It'll be our secret."

She snuggled the bear tightly and fell asleep moments later.

In the early hours of the morning she woke because she had to pee. Entering the bathroom, the lingering smell hit her. It was bitter and sour and unmistakable. She sat down on the toilet. She cried.

* * *

"We'll be back before dark." Jessie yelled through the open door. She was sitting on small wooden bench outside and lacing up her hiking boots. She heard a response, probably from her mother, but not the actual words. She was saved by Robert coming through the door and answering for them both.

"Yes, we promise, Mom. We won't go too close and we won't fall in." This prompted another comment, which he did not answer. Instead, he shook his head and asked Jessie. "Shall we take the usual route?" He smiled, noticing his outdated "usual" perhaps. "Will that be OK for your foot?"

"Absolutely. It's much better now." She stood and moved slowly from one foot to the other. "I've been looking so much forward to our walk." As Robert sat to do his boots, she walked toward a large tree and patted the trunk. "I love this tree. Still standing guard. It hasn't changed much, has it?"

"Not much, I suppose. We are youngsters by tree-reckoning."

"And strangers by now." She said, with a faint smile.

"OK, onwards." He got to his feet and started off down the path. "As long as we don't go too close to the lake."

"She said that?" Jessie laughed. "You'd think we were still seven years old."

"Some things don't change."

"Well, you promised we wouldn't. I didn't."

"You always did that."

"Did what?"

"Found the loophole."

"As I remember it, you didn't mind the misbehaving too much. Especially after we discovered that old rowboat."

"But I felt guilty about it. And a bit scared. You probably didn't. Or did you?"

"Sometimes, maybe. But it was fun, wasn't it?" Jessie smiled, remembering. Then she turned more serious. "Dad seems so much older. I noticed it when he came out to move the car. He shouldn't be driving."

"He's doing fine, for his age. He'll be eighty soon, you know."

"Mom seems to be her usual self. Long-suffering, but still going strong."

"She's that bit younger." He said, ignoring the irony. "Plus I think babysitting Kathy's kids keeps her more active."

"Oh, my God. The horror. In fact, both of them. I just had to get out of there." She sighed. "I thought you said Kathy wasn't coming."

"The homecoming of her dearest big sister was not to be missed, apparently."

Jessie gave him a look. Suddenly, she stopped up. "Right turn maple is gone!" She exclaimed. "Almost gone. How sad. It was such a beautiful tree." She caught up with him. "I would have missed the turn."

"Do you still remember the way?"

"I think so. It's been almost—my God, thirty years. That seems impossible. But it was every summer and fall before that." They moved on, side by side.

"We had some good vacations here, didn't we?"

"We did."

"And it's beautiful right now."

"It is. It's gorgeous." They stopped up and looked around. The forest was quite dense in this area. The trees were in full color display and the air was autumnal, crisp and clear with a hint of decay. After a quick exchange of smiles, they walked on, Robert in front. Jessie was listening to the leaves rustle under their feet.

"So how are you?" Robert finally asked. "Really."

"It's been a tough few weeks. Months."

"Mom shouldn't have said what she did. It's not for her to say."

"I was surprised, frankly. I thought they considered Peter the perfect son-in-law. I was always the errant one."

"They are just concerned. They want you to be happy."

"Anyway, I suppose I did provoke her. With my comment about their marriage being a Gordian knot. But, honestly, I don't understand why they stay together. They're always snapping at each other—when they're not ignoring each other."

"Tumble hitch was a good one, though. You've got to give her that."

"Yes. She's still sharp." Jessie paused. "I hope we'll do better than that, Peter and I. Somehow."

He let some silent steps pass.

"But you don't think he'll be moving over?"

"At first I thought he would, right now I don't. But honestly, I don't know what will happen. He loves his work and he loves the institute in London. I do understand. He has the freedom to take his work in whatever direction he wants, great colleagues and practically no teaching. It might be hard for him to get something comparable here." She paused. "I think he was expecting a job offer from Oak Hill. And when that didn't happen. . ."

"But he's well respected in the field, isn't he?"

"Yes, very. He's done some really important work. It's just that Oak Hill is about new ideas and new approaches. We are primarily hiring scientists just starting out as independent PIs. He. . ." She paused. "Anyway, now he's got personal reasons not to move, as well." This came out with more bitterness than she had intended. Robert stopped up abruptly and turned to face her. She almost walked into him. "No, don't jump to conclusions." She said, looking down. "I'll explain. Keep walking and I'll explain." He did. She continued. "It turns out, to everyone's surprise—mine for sure—that Peter has an adult daughter." Robert turned his head briefly, but she was still looking at the path. "Her name is Tina. She's in London now, but she's from Denmark. A bright and lively girl—or so I've been told."

"So you've been told?"

"Yes. I still haven't met her. Well, not officially. She actually. . ." Jessie stopped talking for a moment. "Never mind. I'll get to that part later. The story is unusual, but not long." She could see from behind that Robert was nodding. "So, this girl first made contact with Peter back in June. She was part of an animal rights group protesting in front of his Institute. She pretended she wanted to interview him about using flies for biological research. It's not a bad idea, actually, considering what those groups say they want. . . But I've looked for it and no interview was ever published or posted." She sighed. "Anyway, they met and talked. Pretty quickly, she came out with the real reason for approaching him: Apparently, she had just found out, from her mother, that Peter was her biological father. To complicate matters, she refused to tell him who her mother was. Peter says he didn't believe her at first, so he insisted on a paternity test. He actually carried it out himself, in his lab, so. . ."

"So you trust the result."

"Yes. I suppose I do. She does seem to be his biological daughter. Arrived in his life, fully formed, at the tender age of twenty-four."

"That is a bit awkward."

"Awkward? Jesus, Robert, it's awful. He claims he has no memory of the girl's mother—or of sleeping with her. Even though he met her again—the mother, I mean—last month. I don't know what to think. It sounds unlikely, doesn't it? Either he has selective amnesia, or it's somehow too embarrassing to talk about. But that's not the most difficult part. He's. . . He's shutting me out. It's like. . . There's this big part of him that I didn't know anything about—and he doesn't want me to know."

"What do you mean?"

"Well, of course, she's someone else's kid. Not mine. But it's not just that. He didn't tell me about her right away—only after they had done the paternity test. I mean, how can you not. . .? We've always. . ." Her voice had become fraught with emotion. She calmed herself, but stopped walking. "It continues the same way. If I ask directly, he'll answer. Sometimes. Sometimes he's incommunicado. Unreachable. And now that I'm here, he has installed her in our house, without asking me. I told him I did not approve, but he. . . I was supposed to go back for a visit last weekend. He doesn't seem to mind that I cancelled. And he claims he doesn't have time to visit me. He doesn't want to leave her, is more like it. It's like I've suddenly become a dispensable bit-character in his life."

"I'm sure that's not true." Robert was facing her, looking concerned. "He's probably just trying to work it all out."

"Right, yes, of course. That's exactly what makes it worse. I'm not even allowed to be upset about it. I'm just supposed to accept it, be understanding, make room for it." She scoffed. He was about to respond but she continued. "I think I would have preferred it if he'd had an affair. Then I could be mad at him and everyone would still be on my side. With this, I can't seem to do anything right. She's the one he's worried about, all the time, even though she is clearly a. . . Oh, fuck. . ."

"What?"

"If I told you, you wouldn't believe me." She walked purposefully on, not seeing his expression. He rushed to catch up.

"Try me."

"So, recently, I found a picture of Tina online. Now, she doesn't have a Facebook page or any other online presence that I can find. Which, in and of itself, is unusual for someone of her generation."

"But it's not unheard of. Some of the youngsters. . ."

"Yes, fine," she interrupted, "but the point is, I recognized her. She was at my gym in London at the same time as me, on the same days as me, for half a year. This started way before she contacted Peter."

"You're sure?"

"Yes, I'm sure. Completely sure. She must have been stalking me."

"It might have been a coincidence. She happened to use the same gym."

"No. It can't have been. I know where she was staying and where her group met. It doesn't fit at all. I told Peter about this. But he won't listen to me, not anymore." She made an angry, dismissive gesture with one hand. They walked on in silence for a short while.

"So you think that it's because of this girl, his biological daughter, that he isn't making an effort with the US move?"

"Yep. Pretty much. And the Oak Hill thing. Both." She stopped. "Let's go down here." She pointed at the overgrown path that would take them to the wooden jetty. "Is it still safe?"

"I have no idea. You first."

"Chicken."

"Loon."

They made their way to the lake after several wrong turns and ongoing comments about where to step and where not to. The jetty was too far gone, even for Jessie. They stood still and looked at the lake instead. The last bit of sunshine caught the trees on the other side. The angle of the light made the foliage glow warm and bright at the same time. The wind had died down, so the reflections were almost perfect—a picture postcard view. They started reminiscing about their childhood adventures and misadventures at the lake. They teased and laughed, but gently. Then they returned to the main path and continued their walk.

"Peter is an only child, isn't he?'

"More or less. His father has children with his second wife, but Peter has never been close to them. He's pretty much estranged from his father, as well, although he's probably a perfectly nice man. I've never met him."

"Really? Never?"

"I've spoken to him on the phone a few times. That's all." She paused. "Peter never forgave him for walking out on his mother. He was very close to her."

"She died quite some time ago, didn't she?"

"Yes, when she was in her early fifties. Breast cancer. It was before my time." Jessie sighed. "No, I do get that part. The daughter, Tina, reminds him a bit of his mother. Peter has no family left on his mother's side, as far as I know. So this new connection does something for him. It's not rocket science."

They walked on.

"Did Peter ever talk about wanting children?" Robert asked, carefully, eyes on the path ahead. "I know you were never interested, but. . ."

"I've asked myself that. Did I miss some clue somewhere along the line? He never said anything. He seemed content with our life, with his work, with me—until Tina came along. I've always been clear about where I stand. So it seems. . ." She didn't finish the sentence.

"I wasn't sure I wanted children, either." Robert said, unexpectedly. "I mean, obviously I love Nick and Bea more than anything. But at the time. . . I suppose I was still hoping to become a serious writer."

"I remember that." She turned and flashed him a quick smile. They were walking side by side again. "When I was in grad school, I read some short stories that you got published. In The New Yorker, was it? And a couple of other places."

"Yes." He smiled back. "I sent them everywhere. A few times, it worked."

"I remember thinking they were wonderful. And different, somehow... I was very impressed."

"It was a good start. I was thrilled. But a few published stories is not enough..." A wry smile. "Certainly not enough to support a family on."

"But then why did you get married?" She stopped and put a hand on his arm. "I'm sorry. That's probably not a nice thing to say. Marion is a good person; I know that. But if you wanted to be free, to write... It must have been pretty obvious that she would want kids, no?"

"It was... You and I weren't close, at the time. So, I never told you, but I was in a dark place for a while... Self-doubt, I suppose. The meaning of life and so on."

"And Marion introduced you to God?" She drew a deep breath. "Sorry, sorry... I just never got that part. It's too foreign for me."

"I know. It is for me, too, actually. No, I never fell for that and she never tried to convince me. That's not the way she is. It was much simpler than that. I fell in love. She became the sun and the moon for me—I couldn't imagine life without her. I just had to make that jump, come what may."

"You're more of a romantic than I thought." She smiled. "I like that." They walked on for a bit. "So, the family part, having children?"

"That's the thing. I just let it happen. I'm afraid I also let myself become a bit of a martyr to it. Like if I always did my duty, I couldn't be faulted. I wore that too tight. To support my family, I focused on the academic path: the papers and conferences, the department, the students, teaching."

"Instead of just writing stories."

"Yeah. But if I'm honest with myself, I was also hiding behind it. Using duty as an excuse."

"For not being a successful writer? If you had been, you'd just be teaching creative writing instead."

"Thanks a lot, Sis."

"Am I wrong? Unless you write a mega-bestseller. Look, you're a serious academic and you're obviously very good at it." He punched her playfully on the upper arm. She rubbed it and continued. "The early promotion, your articles and books. I read the reviews. You're very well respected, admit it." He scowled. "I tried reading the actual books. Well, the first one. The rest look even tougher." She paused. "You got me interested in her stories, though. Munro. I love them. The insights she crams into a few pages."

"She does make everyone else look a bit clumsy, doesn't she?" He smiled. "Anyway, at one point, the whole family with kids thing overwhelmed me. I felt I couldn't breathe, I was stuck, I hadn't actively chosen my life, all the usual stuff: A premature midlife crisis. In the end we separated for a while."

"You did? You and Marion? When? I never knew about this."

"Ten years ago, almost. The kids had just started school. That was another thing. She... well, never mind that. It was just six months. It was too hard with Nick and Bea. I couldn't bear it."

"Don't tell me that you stayed together for the kids? Like Mom and Dad. I always hated that, you know? I know that you're supposed to say the opposite, but I wish they'd had the guts to get divorced—instead of carrying on with their endless bickering and recrimination. His affairs—thinking we'd never know—her stoic standing by. I'm not sure I can forgive either of them. Especially not Mom."

"You're being a bit unfair." Robert said, quietly.

"Well, maybe. It was her choice, obviously, but... But constantly trotting it out as a sacrifice—that she stayed with him, for us, that she gave up a better life, for us—I did hate that. Still do."

"I suppose you're right. We were pawns in their game, Kathy maybe even more so. But we're not like them, Marion and I. We're not at each other's throats."

"I know, I know." She said, head down, with a sigh. "You're very good together, actually. Considerate. Loving. At least that's how it seems. That's why the separation thing surprised me, I guess."

"We never lost the mutual respect—or the love, really. It just changed. But yes, the kids did keep us together, in a way. The extra bond they made between us. Our love for them."

"Divorced parents can love their kids just as much."

"That's not what I mean. It's hard to explain. Our bonds with Nick and Bea are like absolutes: something that will never be in doubt, for the rest of our lives. Giving in to that and letting it rule certain aspects of your life can ground you in a way that leaves the rest of you freer. It sounds strange but..."

"So not absolutes from God?"

"No." He shook his head. "You have her all wrong, you know. Marion's faith is based on choosing to believe in something, not in letting it rule her. I don't believe in it myself, but I do understand the choice element. You should talk to her, sometime, for real. She's not stupid."

"I ..." She started and looked at him briefly, then looked down. "Maybe." They walked on. "You were saying about Nick and Bea," she said a little while later, "that what you feel for them is something absolute." He started to respond but she continued, a bitterness gradually creeping into her voice. "You hear that a lot, about parental love—motherly love, in particular—that it is an absolute, beyond choice. People say that as if it makes it better somehow..." She had picked up a stick, but now threw it away. "Isn't that the weirdest thing? One of the behaviors we most obviously share with animals—fierce protection of our offspring, especially by females—is also considered the most holy and virtuous form of love. When it's really just basic biology. I'd like to think that the forms of love that don't involve propagation of your genes are more noble. But it's the opposite. Sad, in a way."

"Well..." He started, then waited a moment. "That wasn't what I meant, actually. Look, you don't have children."

"Not that again. I don't have kids, so I don't understand. I don't have kids, so I'm missing the most important thing in life." She sounded more disappointed than aggressive.

"Jess. I'm not passing judgment here. I'm just trying to talk... about by own, not always well-considered, decisions and about you and Peter and what's going on there. I'm trying to understand."

"I know. I'm sorry. It's just—I thought—"

"Yes?"

"I suppose I thought. . ." She struggled to work it out. Robert didn't push. "Maybe it's naïve or just plain wrong or. . ." She sighed. "I thought that because our relationship was based solely on choice, not mixed in with responsibility for children, duty and all that, that meant it was stronger. We've always chosen each other, actively. Nothing has been automatic. I mean, yes, we're married. We wanted to show each other and the world that we are each other's primary person, that we are an us. At least that's what I thought. . ." She let the words hang in the air. "It seems to me that our choices reflect who we really are. Active choice, daily attention, daily love—that's what makes a relationship beautiful—not being linked by DNA or duty or whatever."

"Strangely enough you're echoing Marion in some of those words. But never mind. Thinking that active choice is the only thing that really counts is too narrow for my taste. It's too philosophical a distinction."

"But it's. . ." She started.

"To me, it's just common sense." Robert went on. "Choice is important, but other things matter, as well. Givens, you might call them. They are also part of us. And, yes, if you are a parent, and this was where I was trying to go, that is an undeniable given. You'll be torn. Often. Torn between different people and your love for them, your other desires, all kinds of stuff. Or stretched. Stretched is a better word, I suppose. That's when it works. That's what I'm trying for. But I've had years and years of practice. For Peter, this is all new. Your marriage is still all those wonderful things that you have built up, day by day, by choice. But he can be torn, nevertheless. Or if you let him, be stretched."

"You make it sound so simple."

"That was not my intention."

"No?"

"No. I'm still working on it." He smiled, unseen.

"You're also talking about really having children—like you did—children in your life. You watched Nick and Bea grow up, helped them along, first steps, shoelaces, every little thing. All those memories are part of who you are. Knowing you as their father is part of who they are. Peter doesn't know this girl. He's basically a sperm donor."

"That *is* different, I suppose. It's hard for me to imagine."

"It's just DNA. He gave this woman some of his DNA. Why should that change everything?"

"I don't know."

"I don't either. It makes no sense."

They stopped as the path ran into a gravel road sloping downwards to the right.

"This is new." Jessie said.

"Yes, there are a few more cottages down that way now, right on the lake. Let's loop around them." They spent the next ten-fifteen minutes navigating slowly through the forest, keeping the lake and the cottages downwards to the right.

"Someone should have made a new path through here."

"Someone should. We're not here enough. Nick and Bea have been down here occasionally, but never for more than a week. It bores them. They're used to always being connected to their friends, being online. You know. Different times."

"I know." She waded through the fallen leaves, thicker on the ground here. "I loved being here. We were fortunate, weren't we? Dad hated the idea of organized summer camps. Our luck. We got freedom. I don't suppose I ever told him how much I appreciated that." She picked up a red leaf, a yellow one and an orange one. "Beautiful, yes?" He nodded.

"This way." He said and pointed to a half-hidden trail. The light was starting to fade. She dropped the leaves and followed him. They soon found the main path and walked side-by-side again.

"Jess. Can I ask you something?"

"Of course, anything."

"Don't get mad. Promise."

"OK, promise."

"Remember. It's just you and me here."

"Now I'm really getting worried." She looked at him but he was looking straight ahead. "But yes, ask away."

"This business of your not wanting children -"

"I've just never been interested in that kind of life. I never felt any maternal urge."

"I know, but walking here, around the lake, it reminds me. That last time we were here together, in the summer, when we were fourteen. . ."

"Yes. After that year, you deserted me. Every summer, you insisted you had other plans. Friends from school had invited you. You had some odd job. You had summer school. Bad excuses, I thought. But they let you stay behind."

"I know, and I'm sorry."

"You left me with a boring, whiny little sister to babysit. Actually, Kathy was a real horror, like her kids are now. Mom and Dad weren't much better. But they wouldn't let me stay in the city with you. That was so unfair."

"I know. But do you know why I didn't come here any more? In the summer? I didn't even remember myself until I started thinking about it again recently."

"Well, I assumed it was the whole family nightmare. The war of the Aitkins. You managed to escape."

"No, it was you."

"Me?" Jessie stopped up. Robert did as well.

"That year," he said, gently, "your anorexia had become so severe. I saw you getting into the lake. You were like a skeleton, practically. I couldn't look at you. I didn't know how to talk to you."

"Robbie. . ." She started. He looked at her face. Although the light was low, he could see the tears. He continued even more gently.

"I'm sorry, Jess. I just couldn't deal with it. I also think that's why Mom and Dad kept you close. They were worried about you."

"Not that that helped much."

"Well, you didn't end up in the hospital."

"I knew the limit." She said with a mixture of pride and regret. "I knew at what weight the doctor could send me to be force-fed. I stayed one pound above that, always. And I saw my assigned shrink dutifully every week."

"So is this why you never had children? A fear of pregnancy? A fear of putting on weight?"

"No. I got over that long ago. It was a phase. Can't you tell? I'm normal now, not a skeleton by any means."

"I know. You look great, actually. I was just thinking. . ."

"I know I promised not to get mad, but can't you see how insulting this is?"

"Insulting?"

"Peter suggested the same thing and I got really angry." She took a deep breath, in and out. "Think about the subtext. If a woman doesn't want to have children, there has to be a reason, something pathological. . . Because it's just so unnatural. . ." Her voice was running up. "I mean. . ."

"Jess, it's just me."

"Sorry. But this really winds me up."

"I won't mention it again, I promise. I just remembered the state you were in and I remembered how confused I was by it. You freaked me out, you know? The only thing I could do was to look away. And that hurt, too."

"OK. I understand, I guess."

They walked side by side, not speaking. The lake was visible again, but the colors were muted in the dim light, almost gone.

"There might be a connection, though." Jessie said, softly. "Peter made me too angry to think about it this way, but maybe..."

"A connection?"

"Or a common cause. Maybe it's about control." She let the word float for a while between them. "Being in control of yourself. Anorexia is about self-control. I always knew exactly what I was doing. I stayed close to the line, but didn't cross it. I was in control."

"Right."

"And having babies, it's the ultimate giving up control. Pregnancy takes control of your body. Then the baby takes control of your life. It defines you. Whether you become a good mother, or a bad one. No matter what you do, it defines you. And controls you. . ." She drew another breath, but this time slowly, relaxing. She was still walking, still looking straight ahead. "I don't know, Robert. Maybe. I never really thought of it like this before, not consciously. I just knew I didn't want to have kids. I thought I had found a life partner who felt the same, but that's. . . well, let's leave him be for now." She offered a sad smile. "So, it's possible that these things are connected. But it has always been my choice, not some pathology speaking."

"I hear you." He waited for a bit, then added. "Thank you."

"Robbie," she said, a little while later, "can I ask you something?"

"Sure. Anything."

"If you had to choose: Marion or one of the kids—who matters the most? Who would you rescue first from a sinking boat?"

Robert sighed. "You can't ask questions like that, Jess. It's not fair."

"I know." Jessie said, her voice heavy.

A few moments later, Robert answered anyway. "Marion would kill me if I didn't put the kids first." His voice was low and even, as if delivering a simple statement of fact. Jessie's reaction was obscured by the forest darkness.

They turned a corner and could see their cottage at the end of the path. The lights were on inside; it looked warm and inviting. They could not yet hear anyone, adults or children. They slowed their pace.

"So, the new job," Robert said, "I still haven't heard much about that. How is it? Is it what you had hoped?"

"Oh, it's fantastic." Jessie said, her smile immediate. "The place is incredibly dynamic. And it's so interesting to be starting something completely new. I finally feel that I'm making a difference." She paused, drew a full breath. "And I'm good at it. I can tell that I'm really good at it. It sounds a bit much, maybe—"

"No, it sounds great."

"It is. It's the greatest." She took a few more steps. "The greatest."

Chapter 11

"Where did you get them?"

Peter had just closed the front door and was coming down the stairs. Dusk was falling and Tina was standing in semi-darkness near the kitchen area, arms crossed. The only lights that were on were those over the dining table. Two photos were on display: The evidence of his dishonesty.

"Where did you find them?" He asked. He was pretty sure he had left them in a drawer in his bedroom, not on display. She didn't answer. She just waited.

"Your grandmother, Anna, gave them to me." He said. He thought he heard a sharp intake of breath, but couldn't see her expression. She couldn't be shocked, not really, he thought. He walked over to the photos. He liked them even more now, especially the curious toddler in the brightly colored jumpsuit.

"When?"

"I went to Copenhagen to try to find you, back when you went missing." He waited for another question. It didn't come, so he continued. "I was worried about you. And a bit upset, I suppose. You broke off contact without warning and you had made it pretty much impossible find you here." Another pause. He sighed. "But you had given me plenty of clues about your grandparents. So I found them. I asked them if they knew where you were. They didn't. But they seemed very happy to have news of you. Before I left, Anna gave me these photos." He picked them up and looked at them again. "I'm sure they'd love to hear from you directly. They were worried about you." He moved a couple of steps closer to Tina. She took a step backwards.

"But you never told me that you'd been to see them. Mormor and Morfar."

"I was looking for you. I found you. That was the important thing. I didn't tell you about the trip because. . . Well, you obviously wanted to hide your background from me." He paused and looked directly at her face. "I also met Maj, your mother, so. . . "

He paused again. "When you came to live here, I didn't want to upset you. You were not in a good state and I figured that if I told you, you might run off again."

"But since then? I've been here, what? A month?"

"It didn't seem important anymore. Now that—"

"Now that what?"

"Things are different, aren't they? I know who you are. You know who I am. I'm not going to make you do anything you don't want to do." He looked at her face again, close enough now to see the remains of her apprehensiveness. "I think we're past that, don't you?"

"I suppose." She unclenched her arms. "It's been nice, being here." After a moment's thought she added: "How were they, Mormor and Morfar?"

"They seemed fine. They were very nice to me." He smiled briefly. "I understand why you liked visiting with them as a kid. Svend showed me the pen he built for the rabbits. He's got chickens in there now."

"I know. They are kind of awful, aren't they?" She allowed a tiny smile.

"Not exactly cuddly." He said, with a sigh of relief. "No, how about we get some dinner going? I'm starving."

"OK. But I haven't started anything. I was too. . ." She indicated the photos in his hand, shrugged, and turned to fill the electric kettle. Peter switched on the kitchen lights. She seemed to flinch and kept her back to him.

"How about we just go out somewhere?" He said, moving backwards, to behind the counter. "There's that Indian place we found last weekend."

She shook her head slightly, but put the kettle down.

"What did she say about me? My mother?"

"Not so much. A bit about your childhood, your school days and how good you were at languages. We mostly talked about how we met, way back then, and why she hadn't told me about you." He paused. "She didn't say anything about the quarrel you two have, if that's what you're worried about." He went to the fridge and opened it, finding a beer for himself and her carafe of juice. He picked up two glasses. Settling himself on one of the barstools, he indicating another one for her. She accepted the seat and the glass. He spoke again.

"A week or two after my visit, Maj called me. At that point, she told me that she was worried you might be seriously depressed."

"She said that?" Tina's voice ran up, in surprise or disbelief.

"Yes, and when I found you, it seemed like you might be. I wasn't sure of anything. So I let you sleep, kept an eye on you, and you seemed to recover. We haven't really talked about that time properly, the missing weeks."

She poured some juice for herself. He waited.

"At first I stayed with a guy." Her voice was calm again. "I thought he was. . . Well, he turned out to be an idiot. But through him, I met Rimi. You remember Rimi, from the Asian shop I took you to?" Peter nodded and smiled. He had liked being introduced as her father. "She let me stay with her for a while. But I could tell that her parents didn't want me around. Then I was on the streets for a few days. The days were not so bad, actually, but the nights. . . I didn't dare sleep. It was too scary. So I just kept walking and walking. Finally, I went to see Emily. The rest of the Eden group. . . I didn't want to see them again. But Emily is nice. And I was so tired."

"Why didn't you call me? You must have known I would help. Just disappearing like that. . ." He let the words fade out.

"I needed some head-space, I guess. It was all—a bit much—and complicated because I. . ." She didn't finish either. They looked at each other. "I can't believe she said that to you. I'll never. . ."

"What?" I took him a moment to realize she meant Maj. "Tell me what it's all about, won't you? It's better that I know. I won't judge. I promise." He looked at her, knowing he was pleading. "Tell me what went so wrong," he continued, "that the two of you can't talk, so wrong that you're even avoiding your grandparents. Please."

"I'm not avoiding Mormor and Morfar. It's just—her."

"Well, all they got was a few postcards. You didn't call and you didn't have any social media presence so they could follow you. You must have known they would be worried." She looked surprised so he added. "When I got your last name, I looked for you on those sites. My students all. . . Well, they're about the same age as you and they. . ."

"I don't like that crap. Facebook, social media, who needs all that?" He though he heard a bit of a sneer in there. "Besides, they can monitor you even more easily that way. The police, the people in. . ."

"I get it. But never mind. When I found you again, I told Maj and your grandparents. I said that you were fine and that you were living here for now."

"But..."

"I also told them to let you come to them, not to push you."

"Thanks."

"I'd still like for you to tell me. What happened that is so terrible?"

"I'll tell you." She straightened up on the stool. "I'll tell you if you tell me first."

"Tell you what?"

"About your mother, my other grandmother. We had a deal, remember? Is it true that I remind you of her? You said so once."

"You do, yes, in some ways, some gestures. . . It's hard to explain. . . I suppose that's why the result of the DNA test wasn't a big shock to me." He half-smiled, then straightened up. "Right. So what do you want to know? I'll give you three questions. Then you have to tell me what happened between you and your mother. Deal?"

"Deal." She seemed to be thinking hard. "What's her name?"

"Was. Her name was Birthe. Birthe Dahl." He said, slowly.

"What was she like?"

"That's a big question." He paused, looking straight ahead. "She was an academic, a biochemist. That's where she met my father, actually, at university. He went into big Pharma, later on. He left us. My mother stayed at the university and got a position there, mostly teaching. She worked hard, she had to, but she was always there for me. We did all kinds of things together. . ." He paused again. "She wore her hair in a ponytail, like you do. There's that thing you do, when some of the hair comes loose, a reflex. It's like seeing..." He smiled again. "She loved the movies, especially old classics, black and white. I didn't always appreciate them. I thought the were too strange, too artificial and too slow." He shook his head. "I suppose that was more than one answer, actually. She loved me very much, I know that. I knew that back then, as well." He stopped for real. She waited, just in case.

"What's the best thing you remember doing with her?"

"The best thing—now, that's difficult. Let me see..." He thought for a few moments. "I've got it. We went on a trip one summer, just the two of us. It might just have been a week. I was eleven. We rented a canoe and travelled down Gudenåen, all downstream, paddling along. From where I sat, always in the front, I could see the world floating right by, like I was a duck or a swan." He smiled. "We'd talk about the animals and the plants we saw along the way. Sometimes, I'd read out loud. It was The Count of Monte-Christo. I loved that book. I haven't read it since, I don't think. At night, we made campfires and tried to cook by them. We each had our own tiny tent: one blue, one red."

"When did she die?"

"It's been... I was twenty-nine, so twenty-three years ago. You'd have been a baby. She was only fifty-four. Breast cancer. It was..." He went quiet for a while. "But that was four questions, so..." He finally said. "Your turn."

She looked at her glass of greenish juice, still mostly full. She twirled it, drank a sip. Finally, she started talking, her voice low and a bit unsteady. She is so young, Peter thought, with a twinge of protectiveness.

"It was just the two of us. Like you and your mother, I guess. Well, and my grandparents, too. Did you...?" He shook his head briefly, not allowing the distraction. She continued. "She made up these fantastic stories about you. You were an exotic foreign prince, a spy or a diplomat. She didn't know your name or it was a secret code-name. I'm not sure when I stopped believing it." She sighed. "I had a good childhood, I suppose. Normal. Except she was always trying to control things too much. When I was little, it was fine, but later... Every day, she asked me about school, about my friends, about what I was doing, how I was feeling. You said she was worried about me being depressed? She used to be that—depressed, I mean. It must have been when I was little—or before. I don't remember it, but I remember her taking her pills every day and explaining what it was for. Lithium. She said it had helped her a lot." She stopped. He waited. "I lived at home until I got my bachelor's degree. I wanted to move out earlier, but it was hard to find anything. My mother's income was pretty good and she lived in Copenhagen, so I didn't get top priority for student housing. Plus she tried hard to make the apartment my place as well, like we were just sharing. But finally, I was ready to move out. I wanted to take some time off. I wanted to do something real and worthwhile, not just study. I had contact with the Eden group already. So..." She took a deep breath. "Before I left, my mother sat me down, somewhat formally. She said she had to explain something to me. So I could take proper care of myself." Another deep breath. "It turns out she had been dosing me with medication, her medication, the last five or six years. She had told me they were vitamins and supplements and that they were especially important for vegans. Stupidly, had I believed her."

"No prescription? You didn't know?"

"No. I trusted her. I believed her about the vitamins and all that crap." Tina bit her lip. "She said that the dose was very low, but that I should keep on taking it and let her know when I needed more. It was for my own good. She was my mother and a doctor and she knew that I needed it. Blah, blah, blah. . . I didn't know what to say at first. I was so angry."

"She didn't get you to see a psychiatrist first?"

"She said that she had tried to make me talk to someone, years before, but that I'd refused. Things had been a bit crazy for me. That's true. But I was a teenager then. It's normal. . . And no, I hadn't wanted to bare my soul to some judgmental professional. So she drugged me instead."

"Drugged you is a strong term. She shouldn't have done it that way, obviously, but. . ."

"Lithium is a drug, a mind drug." Tina said sharply, letting the anger show. "It makes you become. . . like a different person. For all those years, I didn't get to be myself, because she drugged me. She made me into someone she could handle. I will never forgive her for that. Never."

They let the final words sit for a while.

"I can understand why you are mad at her." Peter said, with slightly strained calm. "She shouldn't have done that."

"No, she shouldn't." Tina almost shouted, tears welling up. "Understatement of the century, I'd say. Those drugs messed with my brain. They changed me. I'll never be the real me. I don't even know who that is, who I really am. And she doesn't know either; she never wanted to know."

"Well, you're still. . . Lithium is just a salt. It shifts the balance of certain signals. It's not like it. . ."

"They don't understand how it works, do they? That doesn't make it any better. Worse, I think. It's a mind-altering drug that no one really understands and my own mother fed it to me for years." She said this with a forcefulness he understood not to contradict.

"So you are not taking it now?"

"Of course not. And I never will. I'll never let myself be controlled by any of that chemical shit. Never. It's all poison."

Peter tried to think. He wished he had someone to talk to, someone who knew what to do with this information. She slipped off her stool and took the glass to the sink. He had to think of something fast or she would disappear up to her room.

"That was when Maj told you about me, wasn't it? Did she say it had something to do with me?"

"No, she. . ." Tina looked puzzled. "No. I made her tell me, before I left. I said that if she didn't, I'd go to the authorities and tell them what she'd done to me, so she'd lose her medical license. I don't even know whether that would actually happen, but the threat worked. She finally told me the truth about you."

"And then you decided to find me?"

"Yes."

They were quiet for a while. Tina leaned on the counter, facing him. Then she looked down at her hands.

"Look, Tina."

"Yes?"

"Let's go find some dinner, shall we? I'll tell you stories about—well—whatever you want."

"Farmor Birthe. Tell me more about her. She'd never do that to you, would she? Try to change who you really are?"

"You are still you, I'm sure you. . ." He started, but felt her resistance returning. "No, I'm pretty sure she wouldn't." He got off his stool. "Let's go. I'll tell you everything you want to know about Farmor."

They both moved quickly toward the door. Peter was not sure whether he was trying to distract her or himself from thinking about Maj's actions. Maybe both.

———

"How could you?" He hissed. He was holding the phone tightly to his ear. His other hand was clenched around a pen and both elbows were planted firmly on the dining room table. As soon as he was sure that Tina was asleep, he had made the call. He didn't care how late it was or what she might have been doing. Perhaps he should have postponed until the morning, he thought while waiting for her to pick up. But he couldn't. He needed to confront her now. The phone was an imperfect instrument for the job, but it was all he had. "How could you put a teenager on powerful psychiatric medication, without her consent and without professional monitoring?" He took a deep breath. "Are you out of your mind?"

"It was a minimal dose." She said, after a while. "I researched it extensively beforehand and knew this dose wouldn't do any harm. And I monitored her very carefully. Christina was doing really well on lithium. You should have seen her at fifteen and sixteen. All these terrible deep holes she got herself into—and the uncontrollable manic periods. She'd disappear all night. I was worried she'd hurt herself."

"It's called being a teenager."

"No, no, no. It was much worse than that." Her words came fast now. "It was really frightening. But I recognized it. And I wasn't going to let her suffer like I did. My parents didn't believe in psychiatrists and they didn't believe in poisoning the brain, as they called it. I guess they had had their chemical misadventures and now wanted everything to be terribly natural. But I was the one who was suffering. Still, I believed them and their hippy delusions for the longest time. I let it continue. Agony, but natural." She scoffed.

"Depressions?"

"Really bad ones—and up periods, occasionally. They were much more fun but also scary. The pregnancy changed everything. I stopped listening to them when they told me I should have an abortion. I was alone with tiny Christina and a post-partum depression was closing in. Hearing my history, my doctor suggested I try lithium. It was a lifesaver. I realized how much better my younger years could have been, had I only known. For Christina, that low dose, just a quarter of mine, it really helped. She stabilized, she was able to study and she did well. Everything was fine."

"But you can't just drug your own child because you think she might suffer from this or that. She should have seen a psychiatrist."

"I tried to make her. She wouldn't. And you seem to forget that I am a doctor. I also know my daughter better than anyone. Certainly much better than you do." She paused, but he didn't respond. "It's always been about what was best for her—always." She paused again, and continued in a softer voice. "I should have told her much earlier. I know that. But I kept putting it off. We were doing well again, but it was fragile. I knew it was fragile. I was too afraid. I. . ."

"You shouldn't have done it. It's as simple as that." Peter interrupted. "She's terrified of any kind of medication now, because you tricked her. What if she really needs something? What am I suppose to do if she gets sick and she. . ."

"What are *you* supposed to do?" The emphasis was heavy and sarcastic. Her voice had lost all the softness of seconds ago, its bitterness fully unleashed. "You are not supposed to do anything except play the nice cardboard cutout daddy that Christina can hang out with until she gets over this. When she does, she'll remember all the good years we've had and she'll be back. You will keep an eye on her, get to know her better if that's what you think you need and, most importantly, treat her well. But never, ever think that you can come here and tell me what my child needs."

"Your child? Our child. And she's not a child any more, she's. . ."

"Oh, stop that stupid posturing. You gave up on her before she was even born. You have no right to come here and. . ."

"Gave up on her? I never knew. . ."

"You knew. You just didn't want to know. You didn't want the mess. . ."

"What are you talking about? You never told me. When we talked in Copenhagen you said that. . ."

"I wanted to see if you'd man up. You didn't. The fact that you didn't made me even more sure that I'd made the right decision back then. You weren't worth it. So I let you go a second time."

"I never. . ."

"Do you have access to your email where you are calling from?"

"Sure."

"I'm going to send you a picture. Call me back when you've looked at it and tell me whether you remember." She paused briefly, then added, before hanging up abruptly: "If you still don't remember anything, then you're the one who needs to have your head examined."

He waited by his laptop for fifteen minutes. A one point, he walked quietly up the stairs to see if Tina's door was open. It wasn't. Now that Tina was aware that he was in contact with Maj, he didn't have to be quite so careful. But he wasn't sure how the phone conversation would have sounded to someone hearing only his side of it. The floorboards creaked under his weight. He went downstairs again, closing the door from the stairs behind him. He poured himself a glass of brandy, sat down and checked the inbox again. Still nothing. Maybe he should call and tell her nothing had come through. Or just let it be and go to bed. No, that would never work. He knew that.

Finally, there it was. The title of the email was "Me being brave". It had no text in it, just an attachment, which he opened. As he had guessed, it was a picture of Maj from back then. She was not hugely pregnant, but very clearly so, especially as she was wearing a tight-fitting stretchy dress that did opposite of hiding the belly-bump. It was striped, as well, the broad horizontal blue and white bands emphasizing her curved contours. She looked so young, so hopeful and so different from the self-assured middle-aged woman he had met in September. She was smiling, but with a hint of trepidation. Behind her were large numbers of badly parked bicycles and, unmistakably, the revolving door entrance to the ugly Panum building. Who had taken this picture, he wondered, in those days before selfies? Had she asked a random passer-by to take it, or perhaps a friend? Judging from her expression and from the fact that she had agreed to pose for the photo, it must have been before she talked to him. Because, yes, she did talk to him. He did remember. Now he did. She had talked to him, wearing exactly this dress, showing off the bump. The picture brought it all back, as she had intended. He felt faint—weak and unwell. Before he lost his nerve, he called Maj back. That was the least he could do, he thought. He was not a coward, not now.

She picked up immediately but didn't say anything. She would know it was him at the other end.

"I got the picture." He left a pause but she didn't fill it. He sighed. "And yes, having seen it, I do remember. You came to see me at the lab. It was right before I left for my postdoc."

"You were not happy to see me."

"I was—what—in shock?"

"You thought I was threatening your future."

"I'm not sure what I thought."

"You suggested it wasn't yours—that she wasn't yours."

"How was I to know? How could I possibly know?"

"Because I told you so." A pause. "I didn't pressure you, did I? I didn't demand you take a paternity test?"

"You didn't." He admitted.

"I left the lab quickly and quietly. No fuss." She paused. He said nothing. "And you never wrote, you never called, you never came to visit. You were gone."

"I was very busy." He knew that was a stupid excuse. "And you didn't exactly invite me to, did you?" Another stupid excuse, he thought. Of course she didn't. She had some pride. "I'm sorry." He finally said. "I should have done better. I am sorry if I hurt you." He left another pause. "But I really had forgotten until now. I wasn't pretending." Pause. "I remember that dress, very clearly."

"I gave the dress away." Her voice was calm now, matter of fact. After a heavy, long-distance silence she continued in the same even tone. "Christina is still angry with me. I suppose I deserve it. But I did what I did to help her, to protect her. I have always done everything I could for her. But no, I did not ask her permission. She would not have given it. I know that. It can be difficult to see what is good for you." She paused. "She will forgive me when she is ready. In the meantime, you can give her a place to stay and you can be there for her. But don't you dare turn her further against me. Or try to make any of this official. If you do, I will tell her that you knew all along, but chose to abandon her because you had more important things to do. And that you couldn't be bothered to check on her all those years in between."

"But, I. . ." He started. It was hopeless. He knew that.

"Don't be stupid, Peter. I'm being generous. Christina is still fragile. I understand that. I won't tell her about the past. I won't jeopardize her wellbeing just to get back at you. You are not important to me. She is. And it is important that she has some stability now, even if it is with you."

"Thank you." He found himself saying.

They made some practical arrangements and rang off.

He stayed in the chair for a long time.

—

When he finally went to bed, his conversations with Maj, then and now, kept bouncing around in his head, as he knew they would. There were two things he just could not understand. Was that really him, the callous young man from back then? And how could he have forgotten so completely?

With the picture on the screen and Maj's voice over the phone, that meeting had come back to him with perfect clarity. It might have been twenty-five years ago, but it was all there, every excruciating detail. The dress, the bump, the lab, her hand resting on his desk, her turning away, her shy smile frozen stiff with shock, his arrogance, all that pathetic self-preservation. She had brought this upon herself, he had reasoned. She had not told him to take precautions. And then she had deliberately waited to tell him until it was too late to make other choices.

He pushed the reruns for more information. They changed a bit each time. Was she really shocked—or just disappointed—or not even that? Was she hoping he would step up or was she courting the inevitable rejection, ready to jump at it as soon as he showed any resistance? Was he offered a choice of any sort? He wasn't sure. As he tried to chip away at these questions, something else happened. Other pictures started crowding in, other memories from that edgy, suspended time.

She had had the operation. He knew that. Afterwards, she had told him everything was fine. They had gotten the cancerous lump, all of it. It was gone and she was fine. He should just go. She insisted on this. And he believed her. But somehow, he must have known, he thought now. He must have known she was putting up a brave front for him. He didn't talk to her doctor, not once. Maybe the doctor wouldn't have said much. Maybe he or she would, if he had insisted. Or at least given him the schedules of checks and tests, some statistics and milestones. He was her only family, after all, her son and an adult. If he had insisted. But he didn't. He left. Crossing the Atlantic seemed like a big thing back then. Of course there were flights, if necessary. But she never said anything. Phone calls were expensive and thus kept short. No Skype, so he didn't see her. It had been his father who had finally gotten in touch and told him he should come home. His father—what a cruel irony that was. You need to come home and say goodbye to your mother, Erik had said, bluntly, brutally, but correctly. No, that wasn't quite it, either. It was worse. There had been no accusation. His father had even apologized for going against his mother's wishes. She had told Erik not to call Peter. Erik didn't think that was right. She had protected him and pushed him away. Or was she protecting herself? When he was told, had he been angry with her for not telling him? Or had he secretly been grateful? He wasn't sure, but he worried that he had been both. They had never talked about it. He hadn't said much at all, if he remembered correctly. He hadn't cried much either, certainly not enough. When he got to the hospital, he had been too shocked by the change in her. A year can do a lot of damage. The funeral, only weeks later, had been the last time Peter set foot in Denmark for a very long time.

It was a major piece of luck that Jessie and he had met only after all this had happened. He understood and appreciated that now. He could not have made a life with someone who had known him in those years. He had needed a fresh start, a fresh attempt at who he wanted to be. As he lay in bed, angrily throwing off the covers to rid himself of the useless torment, the hopelessly delayed self-disgust, he also felt a wave of gratitude for that unexpected mercy. The fresh start.

* * *

Ilana and Carol were back in his office, but everything had now changed. Ilana's face was full of barely suppressed excitement and she seemed unable to sit still. Her laptop was on Peter's desk and all three of them were looking at the data displayed there. Graphs, lots of graphs. All the graphs had error bars; on some, statistical significance was indicated by little asterisks, like prizes. Ilana also had several short videos to show. Most of it was new to Peter and he listened attentively.

"This is the Morris water maze." She said, clicking on a small video icon. "The mice learn to find the platform using these cues." She pointed to markings around the edge of the test arena in the video. "Once they learn where it is, they swim straight to it." The video showed a mouse going in a straight line through the milky water, then stopping. "They don't like swimming." She said, solemnly.

"Yes. I think I may have seen this test before." Peter said, but without irritation.

"And this—" Ilana clicked on another icon and a second video popped up on the screen. "This is the Lego test." She held out a small stack of Lego bricks in multiple colors, neatly fitted into a simple shape. She put it on Peter's desk, clearly enjoying her show-and-tell.

"The two-object recognition test." Carol corrected, with a smile. "It's a novelty-based test. Mice like to explore new objects. We use inert objects like this tower of Lego bricks versus a Falcon tube, for example, and we test how well they remember them."

"Neat." Peter said, looking more closely at the second video. The mouse was moving rapidly, poking its nose close to the Lego tower, again and again. It ignored the other plastic object, close by. "Can you show me how it's done, later on? I've never seen this test before."

"Sure." Ilana said, briskly. Knowing something better than him had to feel good, he figured, especially if she could show it off.

"You know the active and passive avoidance tests, I think." Carol continued. "They are similar to the assays you do with flies."

"Right. Context learning. For the flies, we use odors as cues."

"Ours are spatial cues—or just light and darkness. That adds—well, we can discuss those details later. It's pretty standardized. All of these tests are. I thought, well, we thought that it was worth trying multiple behavioral tests. To make sure the memory defect is general, not something assay-specific."

"Good point."

"I started with the passive avoidance test, so that one has the best statistics." Ilana said. She clicked the video off and the graphs were on display again. She enlarged one.

"Impressive." He pointed at a bar in the graph. "So this is the kinase mutant?"

"Yes." She said. "And that's the control." She pointed at another bar.

"That particular mutant is a hypomorph, we think." Carol said. "This is the null and its control." She pointed at the next set of bars. "Homozygous viable and otherwise normal. Apparently normal, I should say. The phenotype isn't actually stronger, but it's in a different genetic background, so. . ."

"Huifen found the kinase. She noticed the defect when she was testing a whole bunch of mutants." Ilana stated, generously. Carol gave her an approving nod. Peter noticed and was pleased.

"These are my results from last year with the Dummkopf protease inhibitor—or solvent alone." Ilana enlarged another graph.

"I remember. An inhibitor is less clean, of course, but this one is supposed to be quite specific. This was why we were so convinced Numbskull would have the same phenotype. Well—live and learn." He didn't look directly at Ilana. There was no need. She had learned to trust the data in the best way possible. He studied the graphs for a bit longer, noting decay times and effect sizes, trying to compare it to the fly data in his head. "Excellent work, Ilana." He said, sliding his chair back a bit. "Thank you, Carol." He nodded in her direction. "And Huifen, in absentia. You've saved this whole project. And quite possibly my livelihood." He smiled.

"You are most welcome." Carol said, returning his smile.

"This would be a good project to present at the review, don't you think? I know it's only two weeks away, but the data seem firm enough."

"Yes, I agree. The committee likes to see ongoing work—and collaborations."

"So," Peter said, firmly. "Who will do the biochemistry?"

They talked for a while about experiments to be done and initial thoughts about publication. Then Ilana packed up her laptop and left the office in a happy bustle.

"It's so nice to see her energized again." Peter said, once the door was closed.

"Yes, it is." Carol said. She showed no signs of leaving, though. Her expression changed from the benevolent encouragement that had dominated when Ilana was present to something more serious. She fixed Peter with a stern and worried look.

"Hans told me about Jessie's new job. It sounds. . . fascinating."

"It is. It's an interesting approach. They call it Spark publication. Their first "issue" I guess you'd call it, came out online almost a month ago. It created quite a stir. The next one is. . ." He paused. She still hadn't changed her expression. "It's a great opportunity for her." He added, but subdued.

"I guess I haven't been paying that much attention. Actually, it was Lucas or Darya—one of our young PIs—who told me about the Spark thing. They wanted to try it and asked about applying. But what I. . ."

"What you. . .?" He knew what was coming.

"They may not know that Jessie Aitkin is your wife. And that she appears to be working at Oak Hill. In the US, not in London."

"Yes. She is." Dragging his feet, he was.

"So, Peter. Tell me. Hans doesn't seem to know—or else he won't say. Are you two splitting up? Or are you on your way to the US as well?"

"We're not... And I... I'm not sure, Carol." He said, almost pleading. "That's the honest truth." He caught her eye. Her expression softened.

"I'd hate to see you go. You are a terrific scientist and a great colleague, a major asset to the institute."

"Thanks, Carol, I..."

"No, I mean it. This project is just a small part of it. We need people like you here." She sighed. "But speaking as a friend, I understand that it's more complicated than that. Two-career marriages have their challenges, but I always thought the two of you had it worked out. You were both doing well. And you seemed so happy together."

"We were—we are. But yes, it's complicated." He folded his hands together and bent over them slightly, thinking. Carol stayed where she was, waiting. "I did actually send out a few discreet feelers over there." He finally said. "Once it became clear that Oak Hill wasn't going to hire any more senior PIs." He paused. "I haven't told anyone about this. Not even Jessie. I'm not sure why, but... The response was not encouraging." He sighed. "I suppose our work is not really "in" right now—not the flavor of the month—or year. Anyway, as nothing came of it, you may be stuck with me." He added a weak smile.

"I, for one, will be thrilled if you stay. But maybe your feelers were too vague? I get enquiries from Universities over there all the time. Would I be interested in applying for this or that professorship, even chairmanships?" She glanced at him. "We're the same age. And your CV is better than mine. I've just seen them side by side, in the review documents."

"But the last few years..."

"I have the same problem. Of course I work with mice, not flies."

"That makes a difference."

"It does, probably. But it also makes me expensive." She cocked her head. "I'm afraid you just don't have enough X chromosomes."

"Carol..."

"No, I'm serious. A female scientist with your publication record would have job-offers in a heartbeat. I know that. You do too."

"Well." He sighed. "Whatever. That doesn't really help, though, does it?"

"I suppose not. But it makes me so angry, sometimes. My papers are written with the same ink as yours are. Not pink versus baby blue. We should both be judged as scientists. Equally. This preferential treatment crap is demeaning."

"But it's harder for you, earlier on."

"It's hard to manage careers and kids, all at once. Yes. But that's a separate problem. I would have loved better childcare and more hours in the day. I would have loved not feeling bad about going to meetings, bad about not going. William was great, actually, but he never seemed quite so torn."

"I get that, Carol, but it. . ."

"Sorry, sorry. Ignore my usual rant. I wanted to hear about you. What are you going to do?"

"Well, Jessie and I will be meeting up in Madrid next week."

"Somewhere romantic, I hope?"

"Yes. A very nice hotel in the city center. We've stayed there before, years ago. It was great, quite luxurious actually. Oak Hill is paying. So. . ."

"So why not? Enjoy. Have fun. And make Jessie come back to London, won't you? No, scratch that. I don't want to interfere. I just want you both to be happy."

He smiled. "That sounds like a good plan, being happy."

They locked eyes and hands briefly. A moment later, Carol heaved herself out of the chair. She looked ready to take on the world again.

———

He met Gerald on the stairs that same afternoon.

"Peter," he said, stopping up, "do you have a moment?"

"Sure." Peter said. "Now?"

They ascended one floor to Gerald's office. It was modest, considering his position. Peter knew he preferred it that way.

"So, what's up?" Peter said, while Gerald started preparing espressos on his sophisticated machine.

"The review is two weeks away. I understand you'll be away next week, so I just wanted to go over a few things."

They settled in the modern chairs by the small table and quickly talked through the presentations from Peter's section. Peter added a few extra words about Darya's recent findings, explaining why he found the work so exciting. He was planning to highlight it in his overview. He also updated Gerald on his collaboration with Carol.

"Carol told me just now." Gerald said. "It's a perfect story for the review. It shows what we can do with different systems under one roof. And the possible drug-angle. . . Excellent." He smiled. "I know you've been concerned about your output the last couple of years."

"I try to make interesting, complete stories, but that also means. . ."

"I appreciate that you have a small lab and that genetic approaches take time." Gerald nodded slowly as he said this. "This means some years without a top publication. But I have full confidence in your ability to contribute to the field and to the institute for many years to come." He paused briefly. "In fact, I'd like to make you my second-in-command here, if you are interested."

"Wow. Thank you, Gerald. I'm honored..."

"It wouldn't be anything too onerous, I promise, mostly responsibilities internal to the institute. There's simply too much for me to do. In the appointment, I want to reward community spirit as well as scientific excellence. The younger group leaders speak very highly of you as a mentor and a colleague."

"But won't Bill be. . .?"

"Bill has done great science over the years, no doubt about it. But he cannot inspire the younger group leaders—or even the students. As you know, he tends to be a bit contemptuous of anything that's not in his area. No, he's gradually on his way out. He'll be reducing his lab size after the review." Peter thought about what might be driving this. Bill's own choice? The reviewers' written comments? Gerald's opinion? He felt he shouldn't ask. Gerald was still talking. "You don't have to tell me your answer straight away. But if you could give me a hint before the review kicks off, that would be most useful."

"Yes, sure. I. . ." He suddenly remembered Gerald's comment about the collaboration. "What about Carol? She would be terrific as a second."

"She would." Gerald smiled. "And since we have always been straight with one another, I will admit that I did ask her first. But don't take that the wrong way."

"I won't. I respect her way too much for that."

"Good. Carol told me immediately that she did not want the job. Her lab is fairly big and she's very protective of her time. She also does a considerable amount of outside advisory work. We agreed that you would be an excellent choice."

"I will certainly give it serious thought. This place means a lot to me." He stalled for a moment. "What else did Carol tell you? About me."

"She did mention that you might be considering leaving our little paradise."

"It's not that. . ."

"I know. It's complicated. Personal. But I thought, well, we both thought you needed to know how much we appreciate you here."

"Thank you, Gerald. Much appreciated in return." He took a deep breath. "I do love this place."

"I know."

"I mean, with the exception of one or two eminent scientists who don't fully accept the awesome power of genetics." He smiled.

"We all have our crosses to bear, don't we?" Gerald smiled as well. Then he got up from his chair and held out his hand.

———

"It's just so nice to be appreciated." Peter said, his eyes dancing happily atop his huge smile. "It never stops being important, no matter how old you get." He looked at her intently, as if posing a question. After a bit, he continued. "I told Gerald there was one condition, though. From now on, always vegan sandwich choices at the café and only organic dairy products."

"That's good." Tina said. A slow smile suggested she appreciated the gesture. She was sitting by the counter, with a cup in front of her. He moved closer, thinking he might give her a happiness hug, but stopped himself.

"You see, I have this idea for how we could make the students and postdocs interact more. And the lab space. . ."

"Can I visit your lab again?" She asked. He frowned. "I liked doing the DNA stuff with you. Even the flies are kind of interesting. Your student. . ."

"Mihai."

"Yes, Mihai. He's very good at explaining." She paused. He was still thinking. She continued. "That place is such a big part of your life. I'd like to. . ."

"Of course. Sure. We can go tomorrow, or Sunday, if you'd like. It's quieter on the weekend." And probably only Mihai would be in the lab, he thought.

"Yes, please." She smiled and slid down from the stool, cup in hand. "How about Sunday? Tomorrow, it's supposed to be really nice out. We can go on that long walk over to the heath, like we planned."

"Sunday it is." He'd have to tell people about Tina anyway, he thought—and soon. It wasn't fair to Hans that he had to keep it secret. After Madrid, after the review, when everything would be in some kind of order again, that's when he'd tell them. With that decided, he suggested they go out for a celebratory dinner. They discussed the options.

* * *

As expected, the institute was deserted on Sunday morning. Sunday was a day of rest for the hardworking building. Saturday as well, but in a different way, Peter explained. Saturday was the end-of-the-week extra day. Those who came in tended to be relaxed, even if they ended up staying for much of the day—extra coffee breaks, long lunch, random chats. Sunday was even more quiet, but with little spurts of activity in the afternoons and evenings: people getting a jump on a hectic new week to come.

The tall, open space above them ended in giant window-panels in the ceiling. They showed the uniform, gray sky of a dull and dreary day. It had been a good choice to go for the walk yesterday, he realized. It had been one of their best days together so far. They could both talk so much more freely now. He liked that. Everything, well, almost everything, was in the open.

He had kept the temporary entry pass labeled Tina Dahl. She tried it and it still worked. They walked slowly up the wide steps. He was curious if she would remember where his lab was and let her get in front. Once on the second floor, she walked rapidly in the right direction, with a skip every now and then. Two-thirds down the corridor, she turned around and, cocking her head questioningly, pointed toward a glass door. He nodded and caught up with her soon after.

Mihai was in already, working in the wet-lab. He and Tina exchanged greetings, it seemed, but when Peter got closer, Tina moved toward his office door.

"Good morning, Mihai. You're in early on a Sunday." He unlocked his office, while glancing back at Mihai.

"It's not just any bloody Sunday." Mihai smiled, with unconcealed excitement. "How long will you be here for?"

"Not sure. We're just... Why?"

"I'll have the final data from the retest in two hours. Would you like to hear about it? It's looking pretty good." His voice climbed playfully on the last bit.

"Would I? You bet. So you are finally going to stop torturing me and tell me what the gene is?"

"Yes. Now that I'm sure." Mihai smiled. "So can we talk at -" he looked at the clock on the wall "- at noon?"

"Absolutely." Peter said. "I'll be ready."

"You look happy." Tina said, as he entered his office. She had already occupied the resident guest chair and was swirling it around.

"Mihai is finally going to reveal his big result. I just need a little bit more patience—hard as it is." With an expression of mock exasperation, he turned his chair around and clicked on his desktop computer. Checking his email first thing was a reflex. There was nothing new, of course.

"What's this?" Tina was holding up the small Lego tower Ilana had left behind on Friday morning.

"Lego tower." Peter said.

"I see that. But why do you have Lego in your office?"

He agonized, but not for long. It had to be done.

"You know that I work on flies. . ."

"Ye-es." She looked puzzled.

"Well, that's what I've always done and still what we mostly do in the lab, but. . ." He took a deep breath. "But sometimes we have to check whether something we learn in flies is also true in mammals." He searched her face for a moment. Her expression was still a bit puzzled, but not distressed. "Otherwise we won't know whether what we have learned can be used to help people, to develop drugs and so on." He hurried on. "In this case, we've been testing some mice, some mutant mice, to see if they have trouble remembering things, like our fly mutants do. One of the ways we can test this is using Lego blocks." He took the little tower from her, grateful for its playful associations. "We check whether the mice can remember that they've seen this thing before, and that it's different from another plastic thing." He picked up the clear plastic tube lying close by. She looked at the two items, back and forth. "Mice are very curious." He continued. "They like to explore new things, look at them, sniff them, feel them."

"Can I see?"

"What?"

"The curious mice that play with Lego."

"I'm not sure. . . it's. . ."

"I love animals, you know that. I'll be very nice to them, I promise."

"OK." He said. "Why not? We have time."

He took the passkey from the drawer and, on his way out, shouted to Mihai, now in the fly room, that he would be back by noon. Mihai gave him a thumbs-up to show that he had heard.

———

Tina seemed to enjoy handling the mice and watching them explore their little arena. Peter guessed that it reminded her of her days with Pelle and the other rabbits. He used the extra wild type mice that Ilana had shown him. They managed to get them quite well trained in the hour and a half they were in the facility. The assay rooms were set aside from the main holding facility, down a corridor. On this ordinary Sunday morning, they had the rooms all to themselves.

On their way back to the lab, they talked about flies again.

"You train the flies to avoid electric shocks, right?"

"Yes." Peter said, a bit surprised that she remembered.

"But not the mice?"

"Well—we do this Lego test. It's quite good. And a swimming test—it's about remembering where a hidden object is, essentially." He paused, briefly, then hurried ahead. "We also have to do avoidance tests—similar to what we do with flies. It's part of the standard repertoire. But the foot-shocks are very slight." He looked at her face, carefully. "They don't actually hurt."

Tina nodded but did not look back at him.

———

Mihai was ready to talk. But he told Peter he wanted to do it in the fly-room.

"You need to see how the tracking system works, so you know how I found this mutant like a—a needle in a haystack, yes?" Peter nodded. "And because it's so cool." Mihai sat down at his station and turned on the carbon dioxide. He put a few flies on the pad and let Peter have the eyepieces. "There are eleven flies there, all coded on both wings. These guys have all been trained in a standard five-run regimen. Left wing is the same for all of them, right wing has two dots, one line, one dot for ten of them, two dots, two lines for one of them. Agreed?"

"I feel like I'm about to be the victim of a card trick or something." Peter said, while flipping the flies over with a brush. "But yes, agreed."

"Two dots, two lines is our mutant; the other ten are controls. So I'm taking these eleven flies, plus an extra one hundred unmarked flies that have been aged and starved but not trained." He anesthetized a larger batch of flies and put all of the flies, mixed together, into the bottom of his contraption. The flies woke up and started to walk along the side of the lower chamber as he was talking. "This is the reader for the left arm." He pointed to a digital display. "The sensor itself is here." He pointed to a thin glass tube below the display. "Victor's special design. Very cool." He grinned. "The odor is coming from the most distal point of the arm." He pointed to it. "This is the one associated with sugar in the training runs." He indicated another display somewhat dismissively. "That one over there is the right arm reader; you can ignore it for now." He turned to face the central chamber. "But watch this." He slid a lever backwards and the top half of the intake chamber was now open to the flies. They quickly walked or flew into it. A few seconds later, the left display beeped and showed a number and a time point, then it beeped a few more times. The right display also started beeping.

"They are entering the distal arms."

"Right. So we can stop it." Mihai pulled at a lever to seal the two arms off at the same time. He took his carbon dioxide gun to a rubberized opening on the left one. The flies fell to the bottom, anesthetized again. "Now let's look at them." He put the flies on the pad. Peter checked and found two marked flies, one of them with two dots and two lines and three unmarked ones. Mihai pointed to the display. 0011 was first, then two dashes, 0010 and another dash. "Dash means unmarked." He said. Peter had guessed that.

"Very cool indeed." Peter said. "Now I assume you don't start out by individually marking every fly in the mixed populations."

"No, that would be way too much work."

"You start by selecting responders from bigger batches of trained flies and you label individual flies for re-identification."

"Exactly. I take ten or so per batch of two hundred. Then I label them for time trials, redoing it several times. That might even have been your idea, actually."

"Happy to be of assistance." Peter said, for fun but also meaning it. "I probably also suggested you counter-select for flies that enter a neutral arm fast."

"Of course." Mihai collected all the flies and after a short wait, they ran the experiment once more. The left display beeped and displayed a dash, then 0011, followed by several dashes and finally a 0010. "We both know the likelihood of that happening by chance."

"Yes." Peter agreed.

"Now, the most interesting thing here is not just how reliable our double-O eleven is." He added a quick grin at the nickname. "But that I finished training him and his pals two weeks ago." He was holding the open vial from the left side of his sorting contraption in his hand.

"Two weeks? That's really something." Peter said. Just then, the flies in Mihai's tube woke up and one or two flew off. "Careful." Peter almost pounced to cap the tube.

"No worries. I have them stocked."

"Of course. Smart man."

"Early on, I was very careful. The double-O eleven stock is derived from a single male that I tested several times."

"Oh, yes. I remember that. The first result on my desk." Peter fixed Mihai with a careful look, narrowing his eyes. "So the mutant remembers the training for much longer than wild type?"

"Much longer."

"And the phenotype is reproducible, even after outcrossing?"

"Yep." Mihai said. "I just rechecked the fourth generation with PCR. The genotyping was done after I did the final behavioral test, so basically the experiment was done blindly. But it was obvious who was who anyway."

Mihai pulled out a sheet of paper with a graph on it. A downward-slanting line was labeled "wild type", a much shallower line "0011 mutant". Peter studied it. Ten time-points, each with error bars. He was impressed.

"This is fantastic, Mihai. What a phenotype." Peter beamed. "And it's a loss-of-function mutant?"

"Yes, I . . ."

"A true super-memory mutant. As we had hoped." Peter laughed. "Mihai, this could be huge. Great stuff."

"I've done all the controls I could think of." Mihai said. "I've also tested the gene with pan-neuronal RNAi and it goes in the same direction. So that's reassuring."

"Absolutely. Let's go over all that tomorrow, shall we? For now, please, please put me out of my misery and tell me what the gene is. I'm dying to know."

"This one." Mihai pulled out another sheet of paper. It showed a simple schematic drawing: a black line with boxes of different shapes and colors along it. "Not one we know about. It has several domains, mostly protein-protein interaction. And just a CG name."

"Which could be good. It can't have been studied much. But this one, this is a catalytic domain isn't it?" Peter pointed to an oval in the middle.

"Yes. Found in ubiquitin hydrolases. There are quite a few hydrolases in the genome."

"So it might regulate protein turnover."

"Maybe not something we would expect to affect long-term memory?" Mihai asked, carefully.

"It might act at the consolidation phase. We can sort out the time of action with RNAi and Gal80ts, probably. You said the RNAi worked, right?"

"Yes. Not as well but. . ."

"Good, good. So maybe this hydrolase targets transcription factors or chromatin or something." Peter was looking hard at the bench in front of him and tapping it vigorously with his forefinger. He was thinking. "Also, that our protein has all these other domains must be significant. It may be regulated itself, or maybe it's acting in a complex, or it needs to be properly localized, or. . .." He looked up, beaming happily. "There's just so much to figure out here, isn't there?" He paused briefly. "And mammalian homologs?"

"Looks like it. But also not well studied."

"Excellent." He smiled again. "Great work, Mihai. Really. We'll talk more tomorrow, OK?"

"OK." Mihai smiled. "And Thursday? Are we having a joint group meeting this Thursday? Because if we are, I think it's my turn to present."

"Thursday?" Peter frowned. It took him a moment to focus properly on it. "No, not this Thursday. I'll be out of town for a few days." He smiled again, but much more hesitantly. "I'm going to Madrid for a long weekend with my wife."

"Lucky you. Madrid is cool. Have a good time."

"I will." Peter said, with determination.

Chapter 12

The hotel commanded an enviable position, overlooking the plaza. Cream-colored and majestic, it gleamed in the late morning sun. Getting out of the taxi, Jessie recognized the grand entrance with double-pillars on either side of the fancifully decorated black-and-gold doors. A doorman opened one of the doors as she approached. He looked like a Spanish charmer. She smiled at him and said thank you. He returned the smile with a quick appraising look. She immediately understood that he might not, in fact, be a doorman. Just a man, waiting for someone. She shook her head with a hidden second smile, hurried inside and went straight to the reception desk.

Only after she had pushed "5" and the elevator was on the move, did she realize that Peter would most likely be waiting in the room. His flight was scheduled to arrive well before hers. With everything that had happened, they had not seen each other, in person, for more than two months. She should have asked the receptionist, so she'd be prepared. She suddenly felt weak, with a nauseating edge of panic. She leaned against the mirror-clad wall. No, she told herself, not panic, it was just tiredness and jetlag. The flight over had been sleepless. She had decided against the sleeping pill, not wanting to arrive groggy and slow-witted. This seemed to have backfired. Exiting the elevator, she steadied her steps and tried to marshal her thoughts. She had been through this so many times in her head: the necessary discussions, the decisions to make. But the pre-prepared arguments seemed hopelessly far away. She read the room numbers: 502, 504, 506. This was it, room 508. A deep breath and she held the card to the reader.

The lock clicked and she pushed.

"Peter? Are you there? Peter? It's me."

No answer. Jessie's heartbeat slowed to normal and she entered the room. She saw his rolling bag in the corner, opened, with the lid propped up against the wall. She recognized it not so much for the color—it was black—but for something about its contents and how it was packed. An empty plastic bag from Heathrow duty-free suggested there would be Champagne in the fridge. She checked and there was. She smiled at this. His much-loved leather jacket, perfect for London but too warm for this gorgeous day, was slung over a chair. A vase full of exuberant pink roses sat on a small table by the window. He must have bought them on his way from the airport, she thought. There was also a note. She moved closer. The sweet fragrance of the roses was overwhelming.

"Saw your flight was delayed. Gone for a walk. Back soon. Love you."

With today's weather, of course he had gone out. He might have some nervous energy of his own to walk off, as well. She glanced at his bag again. A massive wave of feelings—tenderness, love and longing—washed over her, meeting no real resistance. She put her bag next to his and sat down on the bed, heavily. Tears were threatening. She could just imagine his smile, the one he would have worn buying those roses. She took off her outer clothes, dropped them in a messy pile by the bed and crawled under the covers. Within seconds, she was asleep.

———

The light was softer now. She sensed this as she drifted slowly upwards. There was a new sound in the room, as well. Noticing this brought her all the way to the surface. It was the sound of soft tapping on a keyboard. It was a familiar tapping. She turned her head slowly and saw him sitting by the table, his eyes focused on the screen of his laptop. The roses had been pushed back a bit, but still looked beautiful. A slight frown appeared as he tapped some more. It was his concentration frown, she thought, not a worry-frown. The frown lifted again. Too close to sleep for complicated thoughts, she simply watched him for a while, contentedly. She might even have drifted back to sleep, once or twice. Finally, he noticed.

"You're awake." His warm smile grew slowly but warmly, lovingly. He closed the laptop.

"It seems." She returned a soft and sleepy smile. "Have I been asleep for long?" She asked, yawning luxuriously.

"It's almost four o'clock, so a few hours, yes." He got up from the chair and moved smiling toward the bed, not taking his eyes off her face.

"That's why I feel so much better." She yawned again and stretched, exaggerating for effect. "Good bed, too." She moved back and forth under the sheets, as if burrowing. He sat down next to her and put his hand on her hip, through the sheet. She pulled a warm hand from under the covers and put it on top of his. Then she raised herself up to give him a kiss. "How about you join me?"

"Excellent idea." He started unbuttoning his shirt as she lay back down, smiling sleepily again. She took off her underwear underneath the sheet, clumsily, while watching his familiar, unhurried moves. The shirt came off first, put aside neatly, then the socks, one by one. Finally, he stood and removed his pants, then underpants. He carried the pile of clothes to a chair, nonchalantly showing off his excitement as he did so. He crawled in under the sheet, his skin cooler than hers. She let her body do what it wanted to do, as he seemed to let his. It wasn't very complicated.

———

After they had showered, she looked at her phone for the time. She was dressed again, in fresh clothes. "It's just before five." She said, a bit too loudly. He had come out of the bathroom and was standing right behind her. He had not yet dressed.

"Too early for anything to eat or drink, I think." He said. "Crazy people, crazy place. Early dinner is at nine."

"How about a visit to the Thyssen collection? It's very close by and open until seven. I checked. We can find some Tapas after."

"We could do. We can also go to museums tomorrow or Saturday. We have lots of time." He started putting on the clothes from his neat pile.

"But I'd like to go now." She pouted slightly, hoping that he wouldn't ask why. He didn't. He just smiled. She continued, upbeat. "I was reading Rissa's comments about some of the work at the Thyssen on the way over."

"Rissa?"

"Clarissa, my American gym friend from London. She runs a gallery."

"Yes, of course, I remember her. You've been in touch?"

"Sort of." Her expression clouded over. "Email, so far. But I think we might—well, never mind. . . Let's just go." She put on a determined expression.

"OK, I'm happy to go now, if you want. Should we go to our usual place after?" He asked, and smiled when she looked puzzled. "That little bar on the sharp corner."

"Sure." She smiled. "If we can find it."

"We'll find it." He grinned happily. "I remember exactly where it was." He finished dressing quickly.

———

"So much gold." Jessie said. They were in the middle ages, surrounded by Christian Iconography. "And so completely uninteresting. At least to me." She glanced at Peter, who shrugged and continued strolling straight through the first few rooms. She looked closer at some of the images. They seemed pale and lifeless, despite the gold. She convinced herself that she had tried, if Rissa ever asked.

They moved a few centuries further along and the paintings started coming to life. There were pastoral landscapes, village scenes and people who felt real, whether proud or humble. They studied the little worlds in their splendor and their messiness, giving each image its due, pointing out hidden details or unexpected impressions to each other, their voices low but slightly excited with the fresh pleasure of discovery.

"I really like these." Peter said, indicating two paintings a couple of rooms further along. "The filtered light and the feeling of movement. It works." One was a scene in a park with sparkling water, massive trees and only hints of people. The other showed two children playing.

"And the brushstrokes in these ones." Jessie pointed to two paintings further along the wall. She thought she remembered them from their previous visit. "Everything in this room is so fresh, somehow."

"Funny. I don't know a single one of the artists." He turned to her with a smile. "But that reminds me why I liked this place so much the first time around."

"Which was?"

"Seeing all these unknown artists. It was a revelation. They were trying out all kinds of things way before the big names did. Cezanne got famous for doing something a lot like this, years later. So maybe he took it farther, but it wasn't out of the blue. This place made me question the history of Art that I had been taught: the narrow parade of towering genius."

"Rissa recommended a bunch of pictures for us to look at here. So those in the know do know these artists. We are just amateurs."

"Fair point. They aren't unknown. Maybe un-glorified is a better way to put it." He paused. "It does make you wonder how a few individuals get to be labeled as the most important artists of their time. Why does everyone stand in line for the overcrowded exhibitions of the top names? It's pretty quiet here."

"Big names attract. Not just in art."

"Yeah." He shrugged. "Science, too. A few big names get most of the attention, even more so with the eyes of history. The truth is that there are many bright brains, interesting ideas and neat experiments that matter. And that's a good thing."

"That everyone matters."

"No, not quite that. There's plenty of inconsequential science being done, just like there must be thousands of painters who paint nice pictures but don't contribute anything new. But we know that singular genius is not the real driver of science. There are many significant contributions. Most of them don't get much notice beyond their immediate field. But these small steps are real and important."

"Maybe it's the act of history-writing that distorts it into a few big steps and a few key names? Because we think naming equals knowing? Or maybe it humanizes scientific accomplishments? I don't know."

"It certainly does simplify things." He shrugged. "But no matter. The point is, knowing that you've actually taken one of these small, but real, steps forward and contributed something is a fantastic feeling." He smiled, and added, soberly. "We just have to hope that those who evaluate us see it the same way."

She suddenly remembered the dates. Their institute review was coming up. He had to be thinking of that.

"Are you worried..." But he had already moved on.

"Amazing that those colors, in combination, can give the appearance of flesh, isn't it?" He wasn't expecting an answer, it seemed. He moved on again.

"I wonder how they choose?" She asked. "The artists. How do they choose what to paint? I mean, apart from commissioned portraits and all that."

"Something catches their eye? It could be anything."

"It's hard for me to imagine what this kind of professional life must be like."

"I don't envy artists. Their field—it's all so subjective. I suppose they simply have to be confident that what they are doing is important. Maybe they get that in art school. But if no one else sees it that way, it must be hard to keep going, sometimes."

"Now that sounds heartfelt."

"I meant it about not envying them. Scientists rely on intellectual creativity and an open mind as well—plus hard work and a bit of luck. But there's an objective side to discovery. We know when we've found something significant. It's not just a matter of opinion or taste." He stopped, frowned.

She chuckled. "Well, there is some of that, too."

"Of course." He smiled. "There's some. But also something real, tangible."

They moved on.

"So, your review." She said. "Next week, is it?"

"End of the week, yes. It hasn't been a great period for me, publication-wise."

"If you were an artist, you could claim you're doing your best work, it's just ahead of its time." She said teasingly, and immediately regretted it. He didn't respond. "Peter, you aren't worried about your review, are you? They understand that the work has its ups and downs, they—"

"No, it's not that." He seemed lost in thought, but she couldn't tell if it reflected worry or something else. She was ready to let the subject drop when he spoke up again. "Actually, I've got really interesting stuff to talk about. At least I hope the committee will find it interesting." He smiled. "We've got this new angle on the numbskull pathway in mammals, from the collaboration with Carol. I told you about that, didn't I?"

"Yes, you said. . ."

"So, what I haven't told you about, because it's so new that it's still buzzing around up here—" He indicated the top of his head with a circling finger and rolled his eyes. She recognized the gesture and his excited expression. "—is Mihai's latest finding. His new mutant. It's fantastic. He. . ." Peter looked around, noticing the glances from other visitors. He realized he must have been speaking with too much excitement and too much volume. "Sorry." He said. "It's just. . ."

"I know."

"I'll tell you later, OK?" His eyes sparkled, the smile bursting. "Over a glass of tinto and some jamón?"

"Good idea." She smiled as well, but turned away quickly. She was happy for him, she was. But her face may have shown so much more. She didn't want to answer for that right now.

They moved through the remaining rooms at a moderate speed, never far from one another, but also not very close. They exchanged few comments about the paintings. She guessed that his thoughts were now as far from Art as hers were, but in a rather different direction.

———

"It's the kind of mutant I've always hoped to find." Peter was gesturing with his near-empty flat-bottomed glass, the last drop of red sailing about dangerously. They were still on their first glass. "A mutant with improved memory, and otherwise healthy, suggests that the wild type. . ."

". . .has evolved toward an optimal level of memory retention, not toward maximal memory retention." She filled in, while nodding.

"So you understand why I'm so excited by it?" He put the glass down.

"Absolutely. No doubt about the significant insight or general interest on this one, is there?"

"Nope." He crossed his arms demonstrably.

"If some level of forgetting has actually been selected for during evolution, it must be advantageous, somehow. Do you think this is about optimal adaptability to a variable environment?"

"Probably. Makes sense, doesn't it?"

"It does." She paused. "For human cognition I suppose we know this in a different way. The capacity of the human brain is amazing—but remembering too much seems to drive people bonkers."

"That's mostly about how much we retain initially, I think, rather than duration of long-term memories."

"Might be."

"Anyway, human memory is way too complicated. There's so much filtering and psychological interference. Not remembering something doesn't mean the memory is gone." His voice seemed to dip and get swallowed up in the noisy bar. She leaned toward him. He shook his head and continued, a bit louder. "In humans, it's impossible to sort it all out in a clean way."

"Whereas in the fly. . ." She prompted.

"It's clean *and* simple. Find food and sex, avoid bad stuff." He said, smiling again. "And there's so much we can do in this system. We can find out when it acts, which cells it acts in, where in the cell, which other proteins it talks to, genes acting upstream and downstream—the whole nine yards."

"So you already know what gene this is."

"Yes. *And* that it's a loss-of-function. *And* it has a catalytic domain. . ."

"Ah, I spy the potential for an exam-boosting drug." She widened her eyes.

"A bit premature. But who knows?" He grinned.

"And what is it?"

"It's not one of the old favorites, surprisingly. Maybe even a new pathway. It's. . ." He stopped himself, mid-gesture. In the prolonged pause that followed, her expression went from interested to very curious before withdrawing into an amused understanding. He finished the sentence with a dismissive wave of his left hand. ". . .very suggestive." He seemed to drift off, as if following some internal train of thought.

She waited, giving him time to supply more, but did not ask directly. After a while, a sudden expression of delight appeared on her face.

"Look, Peter. I have a great idea."

"Yes?" He said, slowly.

"The stuff you just told me, it's perfect for an observation. You present the mutant and its phenotype, and the identity of the gene, of course, and then your interpretation. Nothing more is needed. I'm sure the discussion afterwards will be brilliant."

"An observation?"

"That's what we call the primary publication unit in Spark." He still looked puzzled. "The Spark journal. You know, what I've been working on for these past months. The first two issues have been incredibly. . ."

"Oh, yes, of course, but. . ."

". . .well received. Each presentation is followed by half an hour of discussion and everything is transcribed in full. Two weeks later it is published as a fully citable unit, clearly attributed to you." She took a breath. When he didn't respond, she hurried on. "A thoughtful discussion of the possible implications of this could be so interesting, don't you think? I can promise you that we'll have a fantastic group of people for it. Even before the second release, we were overrun with applications and expressions of interest from excellent scientists. We don't have an impact factor yet, but you. . ."

"I'm not sure that's really . . ."

"But this is so perfect for it! I could arrange to get you in very soon. There's one in January that would fit well, I think, it. . ."

"Jessie, hang on a second." He said, somewhat forcefully. "I appreciate your enthusiasm, but. . .."

"Sorry. I didn't mean to. . ."

"It's not. . ." He sighed. "I'm sure this Spark thing is fine but. . . I haven't really thought that much about it." He looked away. After a while, he faced her again and spoke with what seemed to be excessive care. "I think this will be a great story. I'll want to try and publish it in one of the big three, the old-fashioned way. Mihai needs. . . no, I need that. I don't want to give everything we know to the competition before we've had a chance to work it out. I don't want to have the interpretation hijacked by some loudmouth with a different opinion."

"That's not how it. . .." She started, but stopped. After a while, she continued very calmly. "That's fine, of course. It was just. . .."

"I want to tell the story my way."

"Sure. Sorry for. . ."

They stopped talking.

The excitement had completely left his face. Could her over-enthusiastic outburst have done that, she wondered. No, why should it? It was just a suggestion. She was not sure whether to steer him back to the science or to leave the topic. She looked around instead. The tapas bar was quite full now. Everyone was talking, laughing and drinking. They had come early and had a corner table toward the rear. She looked at his face again. He was looking carefully at the bottom of his glass, twirling it in his hand. He looked serious. She waited. They had avoided it for many hours now, but it couldn't go on. Maybe this was a good place for the talk they needed to have. Buffered by everyday people and the lively flow of a language neither of them understood properly.

Peter cleared his throat. She took a deep breath. She feared an appeal to closeness and she feared the opposite.

"Jessie. . . we need to talk—about us."

"We do."

A long pause. Finally, he took both her hands in his and looked carefully at her face. "Jessie. You are my wife, my partner and my friend and I love you." He paused, but it was clear that he wasn't done. "Maybe I haven't shown that quite enough." A slight grimace crossed his face. "I've been thinking about things a lot, recently. Thinking about us. We have had so many good, no, wonderful, years together. We have been there for each other and have taken care of each other." He paused. She was blinking rapidly. "We shouldn't just throw that away. We really shouldn't."

"No." She said, very quietly.

"Life is so much better when we are together." They shared a quick smile and he squeezed her hands briefly before he went on. "But I do understand why you went to

Oak Hill. I do understand that this is important to you. But can't we figure something out? We must be able to solve this."

"Did you miss me?"

"Of course I missed you." He said, with a gentle smile. "Very much."

"I missed you something terrible." She almost choked on the words. "I think I went a bit. . ." She sniffled.

"You should have come home. I knew you were upset, but. . ."

"Why didn't you come over?" Again the words choked her up. "To see me?" She almost couldn't get them out.

"You know I couldn't. I had to. . ." He looked away for a moment. "We're here now, aren't we?"

She looked down at the table and tried to steady her breathing. Finally she managed. She looked up and spoke in a calm and reasonable voice.

"You said that we would go, both of us. You said that it was my turn and that you'd look for something in the US."

"I know, but now there's. . ." He shook his head. "I did look, you know—" he drew a deep breath "—for opportunities. I tried putting out some feelers."

"You did?"

"I didn't tell you because—well, because it was so disappointing. Maybe I didn't really want to face up to it, all the "We really appreciate your work, but. . ." Polite brush-offs."

"Where, when?" Her words came automatically. He returned a look of hurt, then mounting defensiveness. "Never mind." She added quickly. "It doesn't matter."

"It would be crazy to leave the Codon institute right now." He continued, after a bit. "Gerald has even suggested that I. . . Well, they're all being extremely support-ive. They really want me to stay." He looked down at his hands. "Apparently, I've been needing to hear that."

Jessie had developed a frown. "They know about us?" She said.

"Of course they do. They've read about your appointment on the Oak Hill website. They asked me what was going on with us and I told them."

"And what did you say, exactly?"

"I said that this job was a unique opportunity for you and that you had to be at Oak Hill to do it—at least for the time being. I also said that we had not broken up—that we were trying to work out a good solution for both of us." He paused. "We are, aren't we?" There was a hint of desperation is his voice. In a flash, she realized that she loved him for that, for the acknowledged insecurity, right there.

"We are."

"Maybe we just need to do this for a while, you know? Long distance."

"Transatlantic?"

"Well, you have your thing. I. . ."

"Do you really think that is a good idea?"

"It's not ideal. But I don't see any other way right now. Maybe you can work from home some of the time. You know, Skype in."

She looked at him, in disbelief. He looked back, genuinely puzzled. She looked down and shook her head. They were both quiet for a while.

"And what about—" She had to ask, she told herself, she had to. "Tina? Is she still there?"

"Yes, but. . ."

"She was stalking me." Jessie almost cried. "You can't. . ."

"No." He shook his head, slowly, deliberately. "She was just working out. Just like you. I asked her and she said. . ."

"But it can't be. She. . ."

"Don't make a big thing out of this. Please. It was just a coincidence. Surely, you can't let this determine. . . Jessie, please. Be reasonable."

She looked at his face—sincere, loving, hoping, reasonable. She swallowed hard.

"So she's still living in our house?"

"I told you. It's just for a short while—until she's back on her feet. It's complicated."

"But. . ."

"And you are not there. It's really not fair to. . ."

"No. . . I suppose not."

They were quiet again.

"It is our home." He said softly. "Ours. I know that. And I miss you."

Tears were threatening again.

"I love you." He added.

"I love you too."

They held hands again, quietly.

Suddenly, she was feeling light-headed. She needed to move, she thought. "I'll get us two more, OK?" She took the two glasses and stood up abruptly—too abruptly. She almost fainted. "I'll get us some food too." She said. "I think I need something to eat." She walked to the bar and ordered, not bothering to try out her poor Spanish. She felt Peter's gaze on her back as she stood there. It felt good. It felt real and good and safe. She savored it. A few quick looks came from around the bar. She savored that, as well. When she returned, Peter looked calm, relaxed.

She picked up a piece of bread with jamón, squeezed it shut with one hand and took a big bite. She chewed slowly, the satisfaction and relief spreading across her face. "Perfect. Just what I needed."

He picked up another piece of bread, squeezed it as she had done, then used it to toast with hers. "To us." He said.

"To us."

—

Peter was fast asleep, snoring like a contented bear. They had splurged on room service and the bottle of Champagne. She guessed it was well past midnight, but there was still plenty of noise from the plaza below. There was a low constant rumble of traffic and an occasional roar from a motorcycle or a fast car. People were talking and laughing as they walked past. One set of voices engaged in a long and passionate argument before they moved off again. She didn't mind. She didn't mind being awake either, as long as it was in this pleasant, drifty, suspended state where nothing was expected of her and where anything that happened would be in a dream.

A crack between the heavy curtains let in moonlight that fell on Peter's neck and upper back. He had pushed the cover halfway off, as he usually did. She watched his right shoulder rising and falling slowly. The light caught the little curls of hair at his neck. She imagined she could see the sparser hair she knew would be trailing all the way down his back. This wonderfully familiar back. She felt herself pouring her love onto it, freely and generously. It was a profoundly satisfying feeling. She snuggled up to him, aligning her body to his. He let out a new sound, sort of a grunt, registering her move but not really waking. He closed his right arm over hers and she was locked in place. She smiled.

—

They had breakfast in the grand hall. It was the room they had been sitting in all those years ago as surprise window-guests to the fancy wedding reception.

"I think it's even the same table." She said, with a jolly certainty.

"I think we—" he looked at the beginnings of a playful protest and changed course, smiling "—I think you might be right."

Over a slow meal, they recounted what they could remember from that day. They agreed on the bride's dress, but remembered different sets of other details. She told him about the dramatic outfits the younger women wore, the bright colors, daring slits and odd accessories, embellishing where needed. He remembered a few stern, older women and their looks of disapproval. They each contributed a few of the men who had been standing to the side or in the corners. As they had back then, they guessed at their relationships to the two families as well as their secret thoughts. On this morning of silliness, they managed to recreate the whole event to their satisfaction. Or almost—neither of them remembered the groom. They laughed at that. Once their plates had been cleared away, they consulted a map to start planning the day's exploration. The weather looked as promising as yesterday's.

—

Peter had left his mobile in the room. When they went back up before going out for the day, he noticed four missed calls. Hans, Carol, Gerald's office, Hans again.

"What is going on over there?" He said, looking at the phone. The moment of hesitation, of reluctance, was there. But he gave in. "I'd better call back, just to see." He sent Jessie an apologetic look. She shrugged and went to the bathroom, closing the door carefully behind her.

When she came back out, he was still on the phone but now pacing back and forth. His voice was tense.

"Which hospital?"

A short answer came from the other side.

"Could you find out?"

This time, there was a much longer answer from the other side. Peter stiffened as he listened.

"What do you mean, is she involved? Of course she isn't. She quit that stuff months ago."

. . .

"Of course I believe her. What do you think I . . ."

. . .

"Right, I will. But Hans, don't..." He looked over at Jessie, his expression disturbingly unreadable to her. Regret? Fear? Anger? He turned to the window and continued talking. "I'll be there as soon as I can. Just don't—don't do anything before I get there. Please."

. . .

"Tell them that I'm on my way. That's all. I'll get a flight somehow and be there this afternoon."

A few more words were exchanged and he finally took his hand from his ear. He stared hard at the offending device in his hand.

She waited for the words, numb and frozen. Somewhere inside, she knew that this small upset, this unfortunate change of plans, was so much more.

Finally, he looked at her.

"Sorry, Jessie, but I have to go back to the institute immediately. There's been an incident. Ilana, my postdoc, it seems she may have been hurt."

"An incident?"

"A break-in at the Institute. Ilana and Huifen, another postdoc, got caught up in it." He paused. "They probably expected the place to be empty."

"They? A break-in? For what?"

He seemed intolerably far away, standing there, looking at his phone again. He looked back at her, but the gaze was distant, as if he didn't see her. Then his expression grew more intense and determined and his eyes searched her face, as if assessing her. She suddenly felt very cold.

"The Eden group, the animal rights group." He said. "Those idiots. They broke into the animal facility and let out a bunch of mice and rabbits. It seems that Ilana and Huifen were there, working very late or very early. Something happened, I don't know the details."

"But Ilana is hurt? And the other one, Huifen?"

"I don't know. Hans wasn't sure. He was on his way to an emergency staff meeting to hear more." He paused briefly. "I have to go. I have to be there."

"The Eden group." She took a deep breath. "That's the group Tina is part of, isn't it?"

"She was. She hasn't been involved with them since we... started talking. But the leader is a real creep. Alistair. He must be the one who planned this. I'd like to..."

"Peter, come on." She fixed him with a steady gaze. "Honestly. You cannot believe that this is a coincidence. An animal rights group breaks into your institute—" she stopped, just realizing the next fact, "—on the first night that you are away from London. And it just happens to be the group Tina was involved with?"

"I don't know what happened. But I'm sure Tina wasn't involved. Not willingly. She wouldn't do that to me." He looked straight in Jessie's eyes. "You don't know her. You just think..." His anger was sudden, unexpected, and threatened to take control. She saw that. She saw him realizing it, as well. He turned away and started fiddling with his phone again. He held it to his ear and waited. From the other end of the room, she could hear the dial tone.

"Tina, you're home."

He must have realized his mistake as soon as he said it.

Both mistakes.

He kept his back to Jessie. She looked at his back and then out the window, into the unbearably bright light of day.

"Have you heard about. . ."

. . .

"I know. Of course you wouldn't. It's just. . . Well, I'd better come back. I'll be there sometime this afternoon."

. . .

"They need me at the institute. One of the people from my lab may have gotten hurt." He finally turned around to face Jessie, but still spoke into the phone. "Look, I'll see you later, OK?"

. . .

"OK, bye."

There was a long silence. They both seemed unwilling to break it. Peter started moving around the room, collecting his things. Then he picked up the phone again and called the airline. The call went on for a while. He sat down at the table with the naïve, impossible roses and opened his laptop. He was still on the phone as well. Jessie could tell that he had not, actually, forgotten about her. His not looking at her was active, not passive. She sat on the bed, trying to collect her thoughts, and herself, while Peter gave the occasional response over the phone. By the time his phone call ended, she had succeeded.

"Jessie, I'm so sorry. This is awful. It was supposed to be our weekend." He tried to engage her with his eyes. She resisted. "It's just that—with her being there—people thinking she might—I have to. . ." This didn't help. She hardened further.

"I know." She said, very smoothly. "I understand. You have to go back. When is your flight?"

"In just over two hours. I've got to grab a taxi I think, but you. . ."

"I'll be fine. I was going to stay on for the meeting anyway, remember? I'll enjoy the rest of the weekend in the city. The museums and all that." He frowned, probably picking up on the artificiality of her tone. She pretended not to notice. "I'll follow you down." She stood up, a bit wobbly, found her shoes and picked up a room key from the bedside table.

"Jessie, I. . ." He sighed and zipped up the bag. "I'll call you later, OK? And we'll arrange something else soon." She was waiting by the door, stiffly. He picked up his jacket and the handle of the bag. "I'll come over for Thanksgiving. Or another time." His look was imploring but he didn't risk touching her.

"I'll follow you down." She repeated.

———

When she returned, the room was full of icy absence and mocking reproach. She had tried to prepare herself. It didn't help very much. She took a deep breath, stepped inside and closed the door behind her. The click of the lock was loud in the emptiness. The walls threatened to collapse on her. She moved to the bed, found her laptop, opened it and started to look for Rissa's notes. She opened the file, but the words on the screen were not up to the task. They could not help, or even hold her attention. She closed the laptop and looked toward the window, instead. Her gaze slowly moved to the table, to the sad flowers and, finally, to the corner without his bag. The space where he had been standing was like a black hole. She could almost see it. It was tugging at her. She started crying as she felt the useless, mushy weakness claim her. There was no other choice right now. She curled up on the bed and gave in to the darkness.

———

Three hours later, she was walking briskly away from the plaza, toward the Reina Sofia.

Chapter 13

"Tina, are you here?" Peter yelled, as the front door closed behind him. "It's me." No response. "I'm back." The downstairs was empty, unlit. He tried a few steps anyway and looked toward the kitchen. Nothing. He turned around and ran up the stairs. As expected, the door to the guest room was closed. He knocked. "Tina, are you in there?" He was about to try the door when she answered.

"Yes, I'm here." He breathed out, in relief. "But please don't come in right now." She added.

"Are you OK?" He spoke through the closed door.

"I'm still not feeling so good."

"Tina?" Half doubt, half pleading. She had been in her room a lot for the past few days, claiming illness. He had been pretty sure it was also to punish him, after their argument. But he had not challenged her. It was better that she was here, sulking, than out on the streets again.

"I'm scared." She said, sounding it. "I'm afraid the police might come looking for me."

"But you said you weren't part of it."

"I wasn't. But who will believe me?"

"I believe you." He tried the door discreetly. It was locked. "Will you come out and talk to me for a moment? I have to go to the lab soon."

"You won't tell anyone that I'm here, will you?"

"Of course not." He paused. She still didn't open the door. "I promise." Another pause. "But we do have to talk, Tina."

"Yes." He waited. Nothing more came.

"Look, I'll be going now. They're expecting me. But I'll see you tonight, OK? Just stay here, don't go anywhere."

He hurried back down the stairs, noticed his rolling bag and its silent rebuke, moved it aside, and checked his jacket pocket for wallet, keys and mobile phone. If he walked fast, he could make it to the lab in half an hour. He wrote a text to Hans. "I'm back. At your office in 30."

———

He focused his attention on the obstacle course presented by London sidewalks on a busy afternoon. He estimated the speed, direction and most likely behavior of fast walkers like himself, slow tourists with maps, inattentive school kids, mothers with baby carriages, small dogs on retractable leads and endless i-zombies. Constantly reassessing the most efficient path, he ended up weaving his way in a rapid zig-zag. He must have looked the maniac, he thought later, his long legs, his speed and his jacket open to the weather.

It almost worked. He almost managed to keep himself distracted. But the high point of the argument was hard to forget.

"What else haven't you told me?" After a prolonged silent treatment, Tina had finally explained. She was disappointed that he hadn't told her everything about his lab, specifically, that they used mice, and more damning, that they subjected them to electric shocks. For some reason, she had obsessed about that: the occasional, tiny foot-shocks. He hadn't considered how sensitive she was to partial, well-meaning truths. He should have told her earlier.

"Nothing, I just thought. . ."

"You didn't tell me that you'd been to Copenhagen, that you'd spoken to my mother, even. It was only when I found those pictures. . ."

"I was going to. I just. . ."

"So—what else haven't you told me? You didn't just happen to find Mormor and Morfar, did you? You knew."

"No. It was your stories. The clues. . ."

"I bet you knew all along." She stared at him with eyes of pure fury as she said this. "Didn't you?"

"No, I didn't. I was looking. . ."

"Yes. You knew about me. You knew but you ran away, didn't you? I could have had a father."

"I. . ."

He hadn't found an answer.

She had run up the stairs. He heard the door slam.

———

When Peter arrived at the small plaza in front of the institute, he saw just one parked police-car. There was no one in it. Two men in dark green boiler suits were working on removing red spray-paint from the glass front of the building. He could still make out the word "EDEN" in large capital letters, each line of each letter multiply reinforced. The writing that came after had been partially scrubbed off and was now illegible. The animals, if any had made it outside, must have dispersed long ago. But some tension was still in the air, he sensed. The temptation to keep walking was strong. Then he remembered that day, months ago, when his whole world had started tipping. He went inside.

———

"I got here as fast as I could." Peter said as Hans handed him a mug of coffee. Hans reached behind Peter and closed the door to his office.

"It's good that you're here. So, do you want an update?"

"Yes, please. I only know what you told me over the phone."

"Carol went with Huifen and Ilana to the hospital earlier, but she may be back by now. She'll know more than I do."

"They both got hurt?"

"No, only Huifen, I think, but it seems Ilana has been insisting on staying with her. She may be in shock. Ilana, I mean."

"Probably. Is Huifen badly hurt?"

"No, not too bad. A cracked tooth and a cut on the chin, I think. Apparently, there was quite a lot of blood, so it looked worse than it was. They had stayed in the procedure room and locked the door after the confrontation."

"Good choice. If they weren't badly hurt, I mean."

"They might have been too scared to go out, not knowing if the intruders were still in the building. They both spoke to the police this morning. Carol and Gerald were with them for that. Carol insisted on taking to them to the A and E afterwards, so Huifen could get some stitches and have something done about her tooth. And to make sure she didn't have a concussion."

"So what happened?"

"Ilana and Huifen saw two activists, both wearing masks. There might have been more. They broke in—or got in—in the early morning, around four o'clock. It seems no alarms were triggered, so the best guess is that they had a passkey or that they hacked their way in."

"Hacked?"

"It's not impossible." He shrugged. "I've talked to the IT guys to see if they have any indications of a hack. I'm a bit sensitive about such things, you see. It seems not, but they can't be sure. Gerald will be told if they find anything."

"Maybe I should go talk to him?" Peter said, but half-heartedly.

"Not now. He said he'd give another update at the end of the day. Didn't you get the emails?"

"I didn't check. I'm a bit discombobulated."

"Oh, yes. Your weekend with Jessie. Bad timing. Was it. . .?" Hans left the question unfinished and Peter waved it off, dismissively. Hans continued. "Gerald is probably busy with the police, going to go through the CCTV recordings. The police seem to be taking this extremely seriously. I'm a bit surprised. I suppose I shouldn't be, but. . . Anyway, Evelyn and Lucas are helping them. I guess they know the facility better than Gerald does."

"Yes, Evelyn is. . ."

"Lucas is very upset, I can tell you that. It seems many of his animals were lost. At the meeting this morning, he looked like he was going to explode."

"They let out a lot of animals, then?"

"Yes, quite a few. Some were still running around in the corridors this morning. Most had left for the streets of London. Can you imagine?" Hans almost smiled, but stopped himself. "The place was littered with empty cages and poop when I came in—quite the mess—you can still smell it."

"Not really. I. . ."

"I think I'll stick to my computers, thank you very much. They must have propped one of the outer doors open on their way out. The night watchman found it that way when he finally noticed something was amiss. That's another thing—why was he. . .? Well, it's not his fault, I suppose. Maybe the plan was to let the animals out on the street later, so people would be confronted with hordes of white mice and rabbits on their way to work. But they got disturbed, so. . ."

"What about Ilana and Huifen? Why were there here so early? And why didn't they hear the intruders right away?"

"You'll have to ask them. All Gerald told us was that they had been working with rabbits in one of the procedure rooms."

"Rabbits. But Ilana doesn't. . ." Peter said. "Anyway, that's not important. So how, exactly, did Huifen get hurt?"

"According to Ilana, and again this is via Gerald, one of the activists pushed her when she wouldn't let go of the rabbits. She fell onto the corner of a table, hitting her head. So it was bad luck, in a way, but it also means there could be additional charges against the activists."

"Do they know who they are?"

"I don't think so. Not yet. But one would hope that they would give themselves up, given what happened. I think they are more likely to go with Huifen's fall being an accident if they do." Hans paused. "I know you said earlier that Tina. . ."

"Tina wasn't involved. I talked to her."

"Are you sure?"

"She said she wasn't involved. I believe her. She wouldn't do that to me. But some of the other members of that group are quite aggressive."

"You know them?"

"Not really, it's just. . ." He paused. "I might know where they hang out. Or used to. But. . ." He stopped and, after a moment, put his hand on Hans' upper arm. "Hans, you are the only person who knows about Tina—that she is my daughter and that she is staying at my place. Could you keep it to yourself for a bit?" Hans looked uncomfortable. Peter continued. "I'm not asking you to lie, just don't volunteer anything right now. Please. Give me the weekend to sort this out. I'm sure she wasn't involved, not directly, but there could be. . . Well, she might know something."

"The police will talk to all of us on Monday, I think. I don't need to say anything to anyone before that." He looked at Peter, with a shrug. "I haven't even told Alessandra about Tina. It seemed too complicated for a second-hand telling." Another shrug. "Well, anyway, I understand."

"Thanks, man." They both pulled a smile at the borrowed phrase.

"No, I get it." Hans repeated, seemingly to himself. "Daughters." He shook his head. "When Elise says something obnoxious—and I mean, she can be really nasty—if it were anyone else's kid I'd think the parents should—anyway, when I'm just about to lose it, slap her or something—just then, I see her when she was three, having a tantrum, red in the face, a tiny person wrestling with the whole world. And I just want her to feel better. I know this whole mess with me and Marjorie upset her so much more than it upset Liam. He's robust, you know? She's. . ." His voice seemed to fade away, leaving him silent, with a far-off look.

Hans finally roused himself and continued, in a more normal voice. "At the briefing this morning, people were upset, of course, but also puzzled. Why here?

We don't do that much animal work. But apparently, this was not an isolated event. There were three other break-ins last night, all sabotage of animal facilities. One was at a commercial lab. I don't remember the company's name, a large product testing facility somewhere out past the M25. It was like here, pretty much, but a lot more animals were involved. The two others were farther away, high-intensity farms. . ."

"I've heard plenty about those places from Tina. And it worked. I've gone all free-range and organic."

"Well, all four places had the word Eden spray-painted on outside walls. The suggestion seems to be that we were chosen because we're central and visible. Lots of people walk by here. So the writing on the wall. . ." Hans grimaced, the pun probably unintended. "It's probably the same reason they chose this place in June." He glanced at Peter. "Maybe."

"So these stunts are supposed to link us together in peoples' minds? It doesn't seem fair, does it? Conflating serious research with cosmetics and ruthless farming."

"But unfortunately, it seems to be working. At least in the press. Several news outlets have picked up on it already. They talk about all four events together and speculate about "Eden"—who they are and what they want."

"They must have sent their propaganda pamphlets to the press immediately—to help with their speculations."

"Maybe they even sent the pictures. You can look it up online."

"I suppose I'll have to. Later." Peter got up from his chair.

"But, Peter," Hans started to get up as well, "do talk to your girl Tina some more. Be sure you know. . . This might get ugly."

"I will. But first I need to see how Ilana is doing. And Huifen—maybe, if. . ."

"Go talk to Carol. She should know."

———

Peter did not want to face his own lab without knowing how Ilana and Huifen were doing, so he followed Hans' suggestion. Carol met him just outside her office. She seemed to be on the way out and was bristling with purpose.

"Ah, Peter. Good. Hans said you'd be back. Can you come with me? Now?"

"With you where?" Was he supposed to know, he wondered? The short day-and-a-half that he had been away seemed to have made a spectator of him.

"To the hospital to pick up Huifen and Ilana. My car is downstairs in the garage. I know driving in London is crazy, but I figured that our girls need to feel well cared for today. The police don't need them until Monday." She had already started walking toward the elevators. Peter followed. "I'm going to take Huifen to our house to stay for a few days." Carol continued once he had caught up. "She's quite upset, poor thing, and I know she lives alone. The hospital says she's fine to go home, but you never know, do you?" She stopped, slightly out of breath. "Silly girl. Why did she resist?" The last bit was said with affection, not irritation. Carol pushed the button for the elevator. "Anyway, William is at home already. He'll take proper care of her. He's good with fragile people, much better than I am. We can bring Ilana wherever she wants to go afterwards. You and I should come back here. Gerald wants us for another briefing." She stepped into the elevator and held the door open for him.

.

"But should I. . .?"

"Ilana asked for you, specifically." Carol said. "I don't know why." She gave him a quick, penetrating glance and let go of the door. He stepped quickly into the elevator. He guessed what she was thinking and figured it was best to prove her wrong. In any case, going to pick them up was the kindest thing to do. The doors closed. Carol busied herself with her bag.

"I heard Lucas was particularly badly hit by this." Peter said.

"Yes, I'm afraid so. His Alzheimer's mice, all aged and so on. He was a bit hysterical, going on about the review."

"But this won't change anything."

"I know, and I told him that. But he's young and he's angry—not in the mood to see clearly yet. He's after Gerald and the police to do everything they can to find the activists and punish them. To the full extent of the law and all that." Peter drew a deep breath, audibly. Carol waited for him to say something, but he didn't. She continued instead. "No one really wants to listen to him, but they also don't want him going off the rails. Making it worse."

"He is a bit of a hothead." Peter finally said, quietly. "Not from here." He almost added. But neither were they.

The elevator stopped at minus third floor of the parking garage, a place where Peter had never been. Carol led the way. "So, let's get our girls settled, shall we?" She said. A car flashed and beeped close by. "I don't know how much Hans has told you, but it seems Ilana has been a real rock for Huifen, all day. I owe her one."

"Why were they in so early?"

"Huifen has these maternal conditioning experiments where she has to remove the young immediately after birth. This particular mama bunny prefers to give birth at night. So they both stayed overnight. Huifen has been helping Ilana, so. . ."

"Makes sense. And it helps to have company for the all-nighters."

"Yes." She started driving up the spiral ramp. "Sorry that you were dragged away from your weekend. Bad luck."

Peter looked quickly over at Carol's face. It was in profile and revealed nothing except that she was concentrating on driving. Peter stayed quiet. The number of things he did not want to talk about kept on growing. Soon he got busy giving Carol directions from the map on his phone. They didn't have far to go, but the traffic required all of Carol's attention.

———

Once they had managed the confusing layout of the hospital car park, Carol took charge of everything else. She left Huifen and Ilana with Peter, while she checked Huifen's final test results and was briefed on the prescriptions she had been given. Peter heard her use her title freely, but it wasn't even necessary. No bullying, no yelling, she was just being her usual confident self. He had the same title and the same level of seniority, but he did not have her natural authority. He found himself not minding the assisting role at all, just noticing it. So far, there wasn't much to assist with. Huifen looked even more subdued than usual, the bandages on her face accentuating her youth and vulnerability. Ilana did not look obviously distressed, but stayed focused on Huifen until Carol returned. The little group soon set off for the waiting car, Carol in front, the girls after and Peter bringing up the rear.

In the car, they didn't talk much at first. The girls were in the back. Carol glanced in the rearview mirror and asked Ilana where she wanted to go after they had dropped off Huifen. Did she want to go home or was she going to stay with someone? Ilana explained that she lived with her boyfriend, Juan, and that she had called him earlier to let him know what had happened. He had promised to be home by now. Carol was obviously pleased with this, as was Peter. One less worry. Ilana gave them an address and Peter entered it into his phone for later.

—

"I recognized one of them. The one who pushed Huifen." Ilana said.

They had just left Carol and William's house, a terraced house a bit further out than Peter and Jessie's. Peter knew the place from various dinner parties. The house was one story taller and a bit wider than theirs. The top floor was William's music room, he remembered, and the two boys shared the floor below that. Peter had phoned William once they were on their way and he had been waiting at the curb when they pulled up. Peter guessed that Huifen had stayed with them before, as she seemed untroubled by being handed over. William had seemed kind and calm without being overly fussy. Carol had said several things briskly to him, about prescriptions and food and clothes and rooms, but had not seemed particularly concerned when he didn't respond. Then they drove off again. That was when Ilana started talking.

"I didn't want to talk about it in front of Huifen." Ilana explained. "She's still really upset." This was directed at Carol. "She didn't want to be talking to the police this morning."

"Well, she didn't exactly say much, did she? It was all nods and shrugs and looking at the floor. She has to learn that we work with the police here. They're not the enemy."

"I think she. . ."

"But, wait a minute," Carol interrupted, "you didn't say anything about recognizing anyone when we talked to the police."

"I didn't put it together until later on. I told the police about the girl and her red Camper shoes and red leather jacket. I told them about Huifen pulling at her mask and scratching the side of her face."

"Yes, you did. Those were useful details. Well done." Carol said, nodding. She was looking straight ahead. Traffic was moving, but slowly. Peter turned sideways to be able to see Ilana.

"I knew I recognized the voice and the accent when she argued with Huifen. Later on, I remembered from where." Ilana looked straight at Peter. "It was that girl who came to interview you a couple of times, back when they were protesting outside the institute."

"Yes, that was the Eden group as well." Carol said, with satisfaction. "Cowardly bastards, hiding behind masks. Peter," she said and glanced very quickly at him, "you must know who this person is."

"I'm sure Mihai can help identify her." Ilana interrupted. "He talked to her for longer than I did and he never forgets a face. Especially a girl's face."

"You probably have her name written down somewhere, Peter." Carol said. "We'll tell Gerald about this tonight and the police as soon as possible." She glanced at the rearview mirror. "Excellent, Ilana. You pay attention and you keep a cool head. You've also been absolutely wonderful with Huifen today. Thank you."

"You're welcome. I just. . ."

"Now, do I turn here?"

They delivered Ilana safely to her boyfriend moments later. He was also waiting on the sidewalk, summoned, Peter assumed, by Ilana's mobile. Peter and Carol looked at Ilana as she hurried toward him and let herself be folded into a firm embrace. She didn't turn around. They drove off.

"Ilana may have been more affected than she initially let on." Peter said.

"I agree. It's easy to understand." She paused, but only briefly. "Now, Peter, about this girl who interviewed you. What do you know about her?"

"Quite a bit." He admitted.

Then he explained, more or less as he had to Hans. The only thing he left out was where she was staying at the moment. Carol commented along the way with several "Really?" and "Gosh" and even "Oh, my God" but no immediate judgment or commiseration. They finished the conversation sitting in the underground car park back at the Institute.

"So you understand why I'm reluctant to tell Gerald about this?"

"Of course I understand. But that doesn't change anything. It's what needs to be done."

"I called her this morning. She told me she wasn't involved."

"People say all kinds of crap for all kinds of reasons." Carol said, firmly. "Wake up, Peter. It's obvious that she was involved, very involved. She was the one who yelled and pushed Huifen. Ilana wouldn't have made that up."

"No, I don't suppose she would."

"And how long do you think it'll be before they find out which card was used to access the facility? Don't be stupid."

They probably knew already, he realized. His heart sank. He thought of last Sunday and their ill-considered visit to the lab. He remembered every detail now. His access card had been in the drawer. It would almost certainly be gone now. He had had a long conversation with Mihai in the fly room, while Tina had been on her own in his office. Right before that, they had been playing with mice and Lego blocks in the facility. How much of that would be on CCTV? He would look ridiculous. How naïve he had been, how. . .

"You have to get her to give herself up." Carol said, very sensibly. "Call her back, tell her to go to the police and explain who she is and what she has done. That must help, coming forward voluntarily. There's the sabotage, that's just a fact, but with Huifen. . . I'm sure both of them will be willing to say that it was an accident. Or at least that it was not intentional. Maybe they'll just deport her. Deport her and bar her from coming back to the UK, of course. I don't know. Maybe she can help the police find those responsible for the other incidents."

"Some of them were her friends. I don't think she'll want to. . ."

"Then she'll just have to live with the consequences, won't she? That's life."

"But. . . But she's my daughter." Peter said, his voice quivering slightly, making him feel even more foolish.

Carol looked at him long and hard, eyes narrowing. "She carries half of your DNA. That's not exactly the same thing."

———

"I'm sorry to drag you in here twice in one day." Gerald said.

"We'll be civil servants soon, with the number of meetings we're having." Bill quipped. Some laughter followed, but subdued. Peter kept his eyes on Lucas, sitting on the other side of the large conference table. He looked furious. "Sorry." Bill added, unexpectedly. "I know this is serious. Sorry, Gerald."

"Right. I won't take up much of your time. I just wanted to give you a short update. Then you can all go home, take the weekend with your families and let it settle. Firstly, and most importantly, Huifen and Ilana, the two postdocs who were caught in this morning's incident, are both fine." He smiled. "Huifen has been to the hospital with her injuries and has been released as fit. Carol is making sure she is well looked after. Thank you, Carol." He nodded at Carol, who responded in kind.

"Secondly, the police are treating this as a serious crime" he glanced at his notes "under SOCPA, serious organized crime something, section one-four-six. As many of you know, they have taken initial statements and they will be back on Monday to talk to some of you again. So, please, everyone, make sure you are available." He paused. "We are helping the police collect the information they need, but it is too soon to say anything more than that." Peter did not look up, for fear of meeting Carol's or Hans' eyes.

"Thirdly, the rough cleanup done this morning will be followed by a thorough cleaning and full disinfection over the weekend. We aim for a fresh start on Monday. We have an important week ahead of us and we can't let what happened this morning derail that." He paused, looking from face to face. "The review will proceed as planned, starting on Wednesday." A collective sigh was heard in response. A sigh of relief for most, Peter thought. He glanced at Bill, who appeared unaffected and at Carol, who might have wanted to protest Gerald's decision, but stopped herself. He looked away before she could return his gaze.

Gerald flashed a minimal smile and looked around again, ending on Lucas. "Now, I know that some of you have lost valuable animals and I know what that means to your ongoing work. I will talk to each of the affected group leaders individually, on Monday and Tuesday, to discuss how the Institute can best help you deal with what has occurred. Obviously, the review panel will understand, so don't worry about them. We will also be looking into our security measures to determine what needs to be improved. But we will not let activists with extreme views dictate anything about the work we do at our research institute." He stopped and pushed back his chair.

"One last thing." Gerald added, his voice now heavier, slower. He looked very tired, Peter realized. "It appears that the break-in here was coordinated with three other acts of vandalism, targeting animal testing as well as farming. Thankfully, sabotage directed at biomedical research involving animals has been in decline recently, but many of us remember what it was like some years ago. The police are understandably eager to shut down the Eden group before their activities escalate." Gerald finally stood. "Thank you all." He looked around the room in one quick take and left through the back door just as rapidly. Some of the group leaders started on questions, first directed at Gerald's receding back and then, seeing this would not work, at whoever was closest to them. Several moved towards Carol, apparently huddling around her. Peter took the opportunity to leave quietly, grabbing the jacket he had brought along from the back of his chair. There was someone he really needed to talk to.

———

"Tina, open the door." No answer. "It's me. Peter."

He tried the door. It was still locked. "You have to come out and you have to talk to me." He took a deep breath. "Come out or I'll force the door." He heard a noise from inside the room. "It's just me. No one else is here. I promise. But I do need to talk to you."

Finally, she opened the door. She remained inside the room, one hand on the door handle. The light from the hallway caught the side of her face and the angry red scratch marks. She was not trying to hide them. Peter gasped but stifled any other response. He willed himself to be calm and in control. Her initial look of defiance faded quickly to worry, possibly fear.

"Why don't we go downstairs? I'll make you a cup of tea." He turned and started down the stairs, not checking if she would follow. She did.

Once in the kitchen, Peter poured water in the electric kettle and turned it on. He opened the fridge, found a beer, opened it and started drinking from the can. Tina had settled herself on a high stool by the counter.

"What kind of tea would you like?" He said, still not looking at her. A mug was found, then a glass for his beer. He didn't repeat his question, but turned around and looked at her slightly bent head.

"How could you?" His voice was hard and precise, far from its normal careful probing. She looked up for moment, as intended, but her expression was not apologetic. "How could you do something so incredibly stupid?" He continued. "Letting out lab animals, onto city streets. Do you think they'll be happy now? Getting run over by cars and busses or being chased to death by rabid rats?" She bit her lip but didn't respond. "A young woman was hurt," he continued, "badly hurt. She was terrified. They both were. And you just ran away, like cowards." She opened her mouth, but closed it again. "These are my colleagues and my friends. How could you betray my trust in this way? I don't understand. I really don't." His voice had changed from tough to almost pleading, which he did not intend.

"Say something. Damn you." He finally shouted, banging the half-full can on the counter. "You used my card. Plus you were recognized." He moved a step closer.

"But we wore masks."

"Argh," he exclaimed loudly, irritated both at the stupidity of her admission and at his own naiveté. "But why? Tell me. Why in the world did you do such an idiotic thing?"

"I wanted to. . ." she started, looking at her hands. Then she straightened up, looked at his face and continued in a more determined voice "I want to make a difference in this world. I need to do something real."

"Something real? Something stupid and dangerous. And illegal, for sure."

"Animals are being treated cruelly every day. They endure terrible pain and suffering, so humans can have cheaper food and all kinds of other stupid things they think they need to have. It is wrong. Very wrong."

"We've been through all this. I understand how you feel. I've tried to do the right things. But. . ."

"A touch of guilt and better personal shopping habits is not enough. People need to see, to understand. Everyone does."

"But you. . ."

"I am a part of something bigger than myself, something important. We will make people understand and society will change. Real change is never painless." The rehearsed, formulaic nature of her words only frustrated him further.

"But this? Releasing lab animals in central London and wherever? Four counts of writing your silly name on the wall? It is stupid—and pointless. What you did won't stop mega-farms from operating or companies from testing products on animals. All that happens is that the fences get higher, the security gets tougher and you all go to prison. You ruin your life—for nothing."

"It is not nothing." She insisted, angrily. "And I am not stupid. I don't suppose the mice from your precious institute will enjoy their freedom for long. But people will notice. That's why we do it. They will see what we've done and start asking questions. They will see what is going on and then—" she drew a deep breath "—and then, things will really change. We're doing something that matters." The expression of defiance was back on her face.

"But you hurt someone."

"She fell."

"Because you pushed her."

"She deserved it. She was torturing those baby bunnies. I heard them squeal. The mother bunny too. It was torture."

"She was just doing her job."

"That's what the guards at Auschwitz said."

"Oh, for Christ's sake. Not that again. You can't. . . Those friends of yours are truly fucked-up."

"No. They aren't. We aren't. We care and we are not afraid of standing by our convictions." Her eyes were full of fire. "We decided to do something about all the evil stuff that's going on. You just don't get it. Or you do, but you can't be bothered to do anything about it." She made a sound of contempt. "You're all like that, all of you: willfully blind and complacent. You don't want to upset your precious little lives."

"What? All of whom?"

"All of you, with your little middle-class lives, your nice little jobs and nice houses and everything nice. You are so incredibly selfish. Most of your life is over so you don't worry about the future. You don't care about the world and all the suffering that's out there. It's just too inconvenient, isn't it?"

"Wait a minute, Tina. Where is all this coming from? I thought we..." He searched her face. She looked straight back, her expression still contemptuous and firm. Then she turned and fixed her gaze on the far corner of the room. He looked at her profile instead. Behind it they were both reflected in the large panes of glass, their postures rigid. He sighed audibly. "I am no crusader, that is true. But I do something that matters. Lots of other people do, too. We do it without hurting anyone and without landing ourselves in prison." He paused. Her mouth was still pressed shut. "There are better ways of making a difference in the world. Constructive ways. The people you hurt, they do that. They are trying to find cures for diseases. You know that we need research to help improve human..."

"Humans, humans, always humans." She hissed. "Even the smallest advance for humankind is worth killing and torturing thousands of innocent animals for, isn't that right? Box them up in little cages, worse than a prison. Rip away their babies so you can have their milk instead. Brand them, cut them and infect them with diseases. Just to help a few miserable humans. Even though humans are the ones ruining the planet. It makes perfect sense."

"It's how we are." He said, exasperated. "If someone you loved was sick, you'd want to do everything you could for that person. If your child was hungry, you'd want to feed it even if that meant some animal had to die."

"I'm a vegan, remember?"

"But not everyone is."

"Maybe they should be." She turned and looked him straight in the face again. After a moment, the hint of a smile came to her lips. "Anyway, don't give me all that bullshit about ridding the world of disease. I know why you do what you do."

"And why is that?" He asked, genuinely puzzled.

"Because you think it's fun. To you, doing science is interesting and exciting. That's why you do it. You told me so yourself." She smiled more fully now. "See? Totally selfish."

"Come on," he said, "that's just not fair. I love what I do, but that's not a bad thing. The work we do really does contribute..."

The doorbell rang.

The cutting words evaporated and they looked at each other in silence, Peter with surprise turning to worry, Tina with panic. Peter looked for his phone to get the time, but it was in his jacket pocket. It had to be well past eight, or nine. Who would come by at this hour? Hans? Carol? The doorbell rang again. Tina was already on her way up the stairs, quietly but swiftly. When she had closed the guest room door, he opened the front door. A man and a woman were standing outside. They were both in their thirties, he guessed, and looked official in their neutral suits. The woman was petite and very neat, but took control with immediate eye contact. The man was a bit overweight; his eyes were flickering, but his body calm.

"Peter Dahl?" The woman asked.

"Yes."

"We're from the police. I'm DS Taylor and this is DC Morgan." She indicated the man with her head and he nodded, once. "We are looking into the break-in at the

Codon Institute early this morning." She paused. Peter waited. She continued. "You work there, do you not?"

"Yes, I do. I was away yesterday but I came back as soon as I heard." He paused. "One of my postdocs, Ilana Perez, was at the institute when this happened. I wanted to make sure she was OK."

"And is she?"

"Yes. She's fine, thank you. Well, they were a bit. . . She and Huifen. Naturally. A colleague and myself drove them home from the hospital. Gerald Marsden, our head of institute, gave us an update afterwards." He stopped, not sure if he was saying too much or too little.

"Yes, we are liaising closely with Mr. Marsden. We talked to the other relevant staff members as well. All except for you."

"As I said, I've been away."

"We would like to ask you a few questions anyway." She added a minimal smile. "May we come in?" They were still standing on the steps. Someone who looked like Mark, his immediate neighbor, was coming down the sidewalk.

"Yes." Peter said. "Of course. Come in."

He led them to the dining table and offered tea or coffee, while picking up Tina's cup as if to drink.

"No, thank you. We are fine." It was still the woman who did the talking. They sat down by the table, both pulling out notebooks.

"I thought the interviews would be continued on Monday." Peter said. "So I. . ."

"They will. Right now, we just have a couple of questions for you, Mr. Dahl. If you don't mind."

"I don't mind."

"So. You were away, you said."

"Yes, in Madrid." He paused. "With my wife."

"So can you explain why your card was used to access the animal facility at three forty-five this morning?"

He felt himself tense up.

"No—no, I cannot. Ilana, maybe?"

"Ilana Perez told us that she and her colleague entered the facility using their own cards. They were doing some sort of experiment overnight?"

"Yes, I understand Ilana was helping Huifen. I don't know the details."

"So your card?"

"It must have been stolen. I haven't used it for a while." Peter felt the sweat pooling, his eyes roaming. He was no good at this.

"And the red shoes over there—and the jacket," the man said, pointing to the entrance that Peter knew was visible from his seat, "who do they belong to?" The woman sent her colleague a quick look, annoyed perhaps.

Peter considered a lie. His wife. It should have been easy to say. But red Campers? That was so specific. He hesitated for too long.

"Mr. Dahl." The woman said, now speaking softly. "We have reason to believe that your daughter is staying with you at the moment." He didn't respond, just looked at his clasped hands. "We know that she was part of the group that targeted the Codon Institute for legal protest, back in June, and, last night, for illegal entry and sabotage. We know that one of the intruders was wearing red Camper shoes and a red leather jacket. Like those." She pointed. "Now, we still don't know exactly what happened this morning. The young woman who was hurt may have been hurt by accident." She looked at Peter, calmly and directly, with an authority that surprised him.

"We would like to sort this out as quickly and as painlessly as we can. It would be in everyone's best interest if we do." She paused. He remained quiet. "Is your daughter here now?" He didn't answer but he lifted his eyes. He may have looked toward the stairs, or maybe it was the red shoes and red jacket that, only now, caught his eye. The woman got up slowly. She signaled to the man to stay put. "I'll just have a look upstairs." She said. "Is that alright?" He may have nodded.

Moments later, he heard knocking on a door upstairs. It had to be the closed guestroom door. He heard the woman speak but no answer. After a while, he got up as well. He joined the policewoman in front of the door.

"Tina, come out, please. It will be better this way."

———

An hour later, DS Taylor had Tina's story in her notebook and a very contrite-looking young woman only two feet away from her. They were seated at the corner of the dining table and both had cups of tea in front of them. Peter and DC Morgan were sitting across from one another a little further away. Neither of them had said anything during the interview.

"I am so sorry that girl got hurt." Tina repeated. "It was an accident. I was only trying to free the baby rabbits. I love animals, you see." She had not tried to deny what she had done but had kept her version of events simple, a bit naïve.

"We are going to have to charge you with this. Do you understand that?" DS Taylor said. Tina nodded vigorously and wrung her hands. "It is not a minor offence. We take sabotage very seriously." More nodding. "Now, I assume you know about the other incidents that took place in the last twenty-four hours, also claimed by the Eden group." Tina looked over at Peter with a flash of panic. "Do you know who planned these other break-ins?" DS Taylor added, keeping her voice steady. Tina shook her head. Peter felt her agony; she was pleading with him, wordlessly.

"I think we have had enough for tonight." Peter finally said. "My daughter is worn out. She is upset." He paused. "She is obviously very sorry about what happened to Huifen and she is ready to help you."

DS Taylor looked from Tina to Peter and back again. Then she folded up her notebook and turned to address Peter directly. "Fine, but we will need to talk to her first thing Monday morning. At the station."

"We will be there and we will cooperate fully. I promise. Just tell me where we should go and when."

DC Morgan wrote something on a pad and handed it to Peter. The address.

"Fine." DS Taylor repeated, standing up. Tina remained seated. "Now, I am familiar with idealistic youngsters who protest against their parents. But this..." She shook her head, severely, still looking at Tina. "If you cooperate with us, we can help you avoid the worst outcome." She turned to Peter. "So Monday morning at nine? I am holding you responsible. She is not a minor, but she is your daughter."

"Yes, absolutely. We will be there."

A few more assurances and details were exchanged. Finally, the representatives of the law were ready to leave. Peter was, all of the sudden, too tired to stand. They found their own way out.

As soon as the visitors had left, the atmosphere of the room changed completely. Tina's expression shifted from remorseful child to angry youngster. She pushed back her chair, scraping the floor loudly, and disappeared up the stairs in a rush. Peter waited for the flush from the bathroom. He heard it. But Tina didn't come back down immediately. He heard her moving about. In the wait, Peter realized that he was very hungry but also that he didn't have the energy to do anything about it. They should go out, he thought, even if it was late. They both needed a change of scene, some fresh air.

When Tina did come down, she was carrying a sports bag filled to bursting. She left it in the hallway, threw her red jacket across it and pulled out the shoes. Heaving a sigh, very demonstrably, she turned and went down the half flight of stairs to where Peter was sitting, staring at her, as if he couldn't understand what was happening.

"Well, Papa," she said, trying out the word for this one occasion, "I think I shall be going."

"But Tina, if you don't go to the police on Monday, they will. . ."

"They will, what? Not give me the opportunity to rat on my friends?"

Peter blinked, rapidly. She was waiting, impatiently, it seemed. Finally he found his voice again. "But I promised to bring you. They'll hold me responsible."

"So it's about you again, is it?"

"No, that's not what I mean. It's your responsibility to. . ."

"To play by the rules of an uncaring society?"

"Everyone is trying to help you. Don't you see that? This way, you might not even get a criminal record, you can go on. . ."

"Or maybe I can claim diminished responsibility. It's not my fault." She said, mockingly. "My mother drugged me. The drugs made me do it. After-effects."

"I don't know how it works here, but that doesn't seem like a great idea. You'd have to bring Maj into it. . ."

"Don't be an idiot. I'm not serious." She looked at him condescendingly, shook her head, turned around and took the first step upwards. Suddenly, she swirled around to face him again. Her expression bordered on triumphant. "But if I were to do it, she would say exactly what I'd need her to say. She owes me and she knows it. She would do *anything*—" she dragged the word out ridiculously "for her precious daughter. In fact, maybe this is the perfect opportunity to broadcast that particular, pathetic story. Doctor drugs unsuspecting teenage daughter and drives her to crime."

"She'd lose her license."

"She should. Anyway, what do you care? She's just someone you screwed once. Before you ran away."

"I didn't. . . Tina, you don't want to do this. She's your mother. She loves you. I'm sure she would agree that you should cooperate with. . ."

"Of course she would. The two of you. . . You'd actually have made a perfect couple, you know? You both think you know everything and you both manage to screw it all up anyway. You're so incredibly selfish underneath all that. . ." She waved a hand in his general direction, her face showing disgust. "No fucking loyalty and no fucking trust—meet my parents."

"You've got it all wrong."

"Do you really think I don't know what you did?" He stared at her, frozen. "You told the police I was here." She continued. "You promised you wouldn't, but you did. To save yourself."

"No. I'd didn't. . . I. . . I don't know how they knew." His mind was racing, looking for the answer. "I really don't."

"I don't believe you!"

"Come on, Tina, don't do this. This is not you."

"Not me? What the fuck do you know about me? You don't know anything. I despise you."

"No, you don't. You went looking for me and. . ."

"And now you know why." The words and the tone were like a punch to Peter's stomach. He heaved. She looked fierce, but she didn't head for the door immediately. She stood still. After a while, her expression seemed to soften.

"We've gotten close." He said, in a low voice. "I know everything is difficult right now, but I am trying to help you. I have come to—" he breathed in, then out "—to really care about you."

Her face scrunched up, all the confidence and hauteur gone in an instant. A tear or two might have run down over the red scratches on her cheek. He couldn't actually tell, from that distance.

"I just want to make a difference," she said, "a real difference—not just be another sad, safe little life."

"I know you do." Peter said, getting up. He considered moving closer, folding protective arms around her and her letting her cry on his shoulder. That, he thought, is what a real parent would do. He considered it for quite some time. But he didn't actually move.

Finally, she picked up her bag and opened the door to the outside. She did not turn around and she did not say goodbye.

Chapter 14

"So would you say that the Spark magazine—or journal—is a success?"

"Journal. Or just Spark. Yes, absolutely. For the first issue, we had as many online views as the top journals in the area. Since then, traffic has more than doubled. And the readers are not just browsing headlines; they click through to view the presentations and discussions. Importantly, scientists are also eager to contribute, both well-known scientists and younger ones. We are being flooded with applications and suggestions." She drew a quick breath. "So yes, we think that Spark is a success. We also expect this to be reflected in the impact factor, when we get one."

"And this factor is important?" The interviewer asked. She had given Jessie her name when they started, but very unusually for her, Jessie had forgotten it. Maybe the bright lights of the studio had distracted her.

"It is. In a way." Jessie said. She was not asked to expand—and didn't.

The interviewer consulted her tablet again. "You described the format for Spark earlier in the program and explained how it differs from traditional formats. Has the new format been welcomed by the scientific community?"

"Some love it, others are more skeptical. But that's good. Scientists are supposed to be critical. We also don't expect Spark to replace traditional journals, merely to complement them. They serve different purposes. For example—" The interviewer's eyes were glazing over again. This was Jessie's cue to cut back. "Well, examples get too specific. But let's just say the Spark format has inspired a lot of discussion."

"Discussion about formats?"

"Well, yes, that too—but I was thinking of scientific discussions. Promoting that was our original purpose."

"So you are talking about the comments published alongside the observations?" The interviewer sounded proud to have gotten this part.

"That first, yes." Jessie tried not to show her frustration. This interview was a bad idea. "And, after that, discussion amongst readers triggered by what they have read. We've been collecting user feedback and it's clear that there's a lot of follow-on from Spark publications, for example at journal clubs. Scientists and students will. . ."

"Let's talk a bit about you and your career, Jessie." The interviewer broke in. Jessie felt a tightening inside and knew her face would be going blank. "Before Oak Hill, you had worked your way to up editor-in-chief of a top ranked scientific journal. This is a very prestigious job, I understand. What made you leave? A risky move."

"Getting this opportunity to try out my idea was a dream come true. I just had to do it."

"So Spark was your idea?"

"Originally, yes. But Anthony Swift and I have worked together to shape it into something that actually works. It is an Oak Hill initiative."

"And are you keen to continue with it?"

"Yes, absolutely"

"Even if your husband can't join you?" The interviewer added a saccharine smile and leaned slightly forward.

Jessie was startled, as intended, but her reaction on camera was minimal. "I don't think my private life is relevant in this context." She replied coolly. Then she looked the interviewer directly in the eyes and waited. It took only two seconds. The interviewer broke eye contact first and looked down, possibly at her tablet.

"So, what can you tell us about the future directions. . ."

———

"I'm impressed." Robert said, taking her arm in his. "You handled her question about Peter like a pro. You shut her right up." He gave her arm a quick squeeze. They were both wearing gloves, scarves and long coats, but the icy January wind of New York's long avenues cut right through it. Tiny snowflakes were being whipped sideways.

"She shouldn't have asked. It was sleazy." Jessie said, looking at him. He returned the gaze only briefly. Then they both looked straight ahead. She felt the miniscule ice particles sting on her not-yet-numb cheeks. "It's only one more block, thank God." She added. "Are you sure you have time?"

"Absolutely. My plane leaves in four hours. Unless O'Hare is under too much snow, in which case it will be later. So, no worries."

"They must be used to this, in Chicago." Jessie's face showed she wasn't.

"We are." He said and wrapped an arm around her shoulder. They walked like this, companionably but awkwardly, until they reached her hotel.

———

"Thanks so much for coming, Robert." She smiled, her cheeks still red from the short exposure. She had placed her scarf, coat and gloves on the barstool to her right. Robert's were on top.

"I wouldn't miss it for the world. My favorite Sis on TV." She didn't protest the nickname this time. "And as I said, I had a layover anyway."

"My jet-setting brother."

"Hardly. And look who's talking. You're always off somewhere—I can't keep up anymore."

"There's a lot of networking and promotion to do at this stage. When I'm not running or processing a meeting, I do that. Tony, he—well, he doesn't travel much. His daughter, she—" Jessie let the comment fade. Robert had never met Tony. She smiled, a bit hesitantly, and touched his sleeve. "So, how is everyone? Marion and the kids? Mom and Dad?"

"We missed you at Christmas."

"It's not really my thing." She tried to look casual. He shook his head slowly, knowing better. "OK, so just not this year." She continued. "I couldn't handle all the questions. Mom and Dad, you know?"

"They only want. . ."

". . .me to be happy, I know." She took a sip of her virgin Mary and looked away, focusing hard on something, anything, to avoid well-meaning questions. She noticed the bright green patches all around—small square stools, pillows on the sofas, on the wall. They were in the lobby bar of a fancy, ultra-modern hotel. Her room was black and purple, she remembered, small but serviceable. She could have taken the train back. But this way, she could head straight out of JFK in the morning for the next leg of the trip. The flight was at nine, so. . .

"I'm surprised they would have such a science-heavy piece on that show." Robert's words interrupted her thoughts. "Positively surprised, but. . ."

"It wasn't even really about science, was it?" She smiled at him, warmly, gratefully. "No exciting new planet or dinosaur or cure to present. I'm afraid it was rather dull. But, yes, it was a surprise. We couldn't really say no."

"Of course not. It's good exposure."

"I'm not sure. Maybe they'll just cut the whole interview. Lack of general interest—since I didn't give them the saucy personal angle."

"You did brilliantly." He beamed. "In my completely unbiased opinion."

"Thanks." She touched his sleeve again, squeezed slightly. "So tell what everyone is up to these days. Nick, has he decided on a college yet? I suppose he must have. And Bea, is she any better?"

The time passed all too quickly.

—

Two hours later, she was pounding on the treadmill, thanking the universe for hotel gyms and for the healing power of time. Her foot was completely back to normal. She could run again. It was dark out and the colorful lights of the city, blurred by drops of former snowflakes, dominated the view. The long rows of slow-moving headlights and taillight were strangely mesmerizing. She tried to work out what an extended patch of blinking lights was all about and to guess what the smudged neon signs might be hoping to sell. After a while, noises in the room filtered through to her, despite her own loud pounding. She had forgotten her earphones. Someone turned on the TV. She tried not to get irritated. They were allowed to do this, she knew, and the sound level was moderate. A talk show was on. She tried to focus on the cityscape again but finally had to accept where her thoughts were going.

The first part of her internal review was predictable. She went through the afternoon's interview in her mind, fast-forwarding through smooth parts and dwelling, with some regret but none too deep, on the parts she could have done better.

What came after was less predictable. Maybe it was being in a new gym that did it. Or being in New York. She found herself thinking about Rissa. She felt bad about not coming to see her in New York in December. It had not been impossible, time-wise, but she had pretended it was. Rissa had been her friend for so long. Couldn't that be salvaged? If she was honest with herself, she wasn't sure. Everything to do with those years seemed like a parenthesis now. It was such a large chunk of her life, but still, that's how it felt to her: A long parenthesis, but a parenthesis nonetheless. Maybe it was best that it stayed that way. She let herself think of Peter. Her feet and legs pounded on, keeping her steady.

They had not spoken since November. At some point, she knew, they would have to. But not yet. In the beginning, he had called several times a day. She had assumed that he had worked it out and was calling to yell at her. It had to be her. Who else had known that Tina was staying at their house and that she might have been involved in the break-in? No one, she had thought. But she had been wrong about that, apparently. Hans had known. But Hans had kept quiet about it. He had not called the police. Peter told her that, later on, in that strange, final email. Better not to think about that email, she reminded herself. It hurt too much.

She should have known he wouldn't forgive her.

She had known he wouldn't forgive her.

She ran on.

Tina had taken advantage of him. Had she known which buttons to push—or had she just been lucky? She had wormed her way into their lives and she had taken advantage. That was not right, genetic offspring or not. Her claim to him was unearned. But he would have kept on protecting her, Jessie was sure of that. He would have kept on lying to himself and the world, pretending he had a noble reason for it. No. She had to do what she did. This way, he still had a chance to do what he loved to do and to move on with his life. He still had his job and his lab. His science was going well again—some new and exciting findings—he had told her in Madrid, hadn't he? Yes, he had. Well, that was good. Getting a new job would be difficult for him, of course. But he was secure at the institute. She had made sure of that when she explained to Gerald how Peter had been tricked. She had told him about the stalking as well—so he would understand what Peter had been up against. The institute would remain loyal, Gerald had promised her, if at all possible. As long as there were no formal charges against him. Tina's disappearance didn't help, of course. She had taken the easy way out. Unsurprising, thought Jessie. The press, intrigued by the bizarre father-daughter story, was not making it easy for him, though. The press. She had not told the press. That was not on her. But maybe it helped him to blame her for everything.

It was OK. She was strong again. She could take it.

Part II
The Essay

Narrative and Its Uses in Science

Everyone likes a good story. A good story can be short: an anecdote, a vignette of a real life happening that reveals a person's essence. It can have a moral, be humorous, or both. A good story can be long: a thoughtful novel read over many disappearing hours or the meandering story my grandfather would add to every evening of our summer vacation. A story has characters that we come to care about or at least to be curious about. It has a plot, one event following another with a logic that is clear or that we hope to divine along the way. A good story, if not too short, will transport us to somewhere else. A story may be true or fictional. In either case, if told well, it will stick with us. Stories have probably been around for as long as we have used language.

What I would like to discuss in the following is one particular type of storytelling, the scientific narrative, and specifically its use in the natural sciences. Although the scientific narrative may employ personal narratives, its primary aim is not to describe the internal or external lives of its characters. Its primary aim is to describe an aspect of science, usually one or more scientific discoveries. The scientific narrative might describe how these discoveries were made or what they actually mean, or both. Scientific narrative is clearly very useful in communicating about science to the general public, but it is also used—surprisingly widely—by scientists communicating about scientific discoveries to other practitioners in their field.

Communication of science to the public is obviously important in a democratic society. It serves to disseminate knowledge and allows informed decision-making about societal issues that have a scientific component. Environmental issues such as global warming and health-related issues such as prevention and treatment of specific diseases are good examples. Most of this type of communication is done by specialized science journalists, who often combine some science background with good story-telling skills. Practicing scientists may also contribute directly in the form of public lectures or articles and books of popular science. A primary concern for both groups is how to communicate something that may be challenging, even a bit dry, in an effective and engaging way. This is where storytelling can be very useful (Dahlstrom 2014). Narrative seems to have privileged status in human

© Springer Nature Switzerland AG 2018

P. Rørth, *Tumble Hitch*, https://doi.org/10.1007/978-3-319-97364-7_2

cognition, a bit like language and face recognition. People generally react well to stories, as opposed to simple listing or explanation of data. They are more likely to listen and pay attention. Stories also provide a framework that makes it easier to retain information.

How is this done? The story could be framed around the journalist's own experience. For example, if reporting on the impact of broad screening for colon cancer, the writer could start by relating the experience of using a home-screening kit. Readers will be able to imagine the slightly awkward event, the strange moment of sending a sample off, the wait for the result and for the possibility of more invasive screening. Along the way, we would get some numbers, statistics and explanations. There could be a family member who has had colon cancer, adding speculation and worry about genetic components to the disease. And so on. If reporting on exploration of Mars, the story could follow a scientist trying to test a whimsical theory with a device on board. There would be the excited wait for the measurement to be made, missing sleep and dinners, possibly a worrying mishap, a correction and finally the data arriving, the theory getting marked up or down. Alternatively, the story could follow a robot as it is exploring this mysterious planet: the landing, moving around and "seeing" what is there, behind a hill, down a crater. Characters in stories need not be human. We just have to be able to relate to them. As in the colon cancer story, numbers and facts would be imbedded in the narrative. Not too many or too heavy, but enough that the reader is better informed after reading it.

Because a scientific narrative aims to illuminate science, its accuracy and proper representation of the underlying data is important. But given the power of a well-told story to engage an audience, using this form is often a good choice for general communication. Unlike communication within science, the attention of the audience, the general public, cannot be taken for granted.

Communication Within a Scientific Field

How does narrative come into play in the presentation of new results within the scientific community? The standard mechanism of presentation is to publish a paper in a scientific journal or to present results in oral form at a conference. In the following, I will focus specifically on scientific papers based on experiments and observations. One paper usually contains a cluster of related results. The choice of journal in which a paper is published provides context on the sub-area of science being explored and, in most fields, also indicates the perceived significance of the findings presented. A commonly used format starts with an abstract or summary, followed by a more in depth introduction, which describes the state of the art for the subject under study. Next comes a methods section with exact details on how experiments were performed, followed by a results section in which all the relevant outcomes from the experiments are show. Finally, there will be a discussion in which the authors interpret their results and speculate about their significance. In the original, classical version of this format, the results are simply listed or very briefly described, largely by referring to the tables and graphs that make up the figures. In biological sciences, the field I am most familiar with, results and whole papers are now—more or less universally—presented as narratives. Biologists have become so used to the narrative form that we use it, teach it and read it without reflecting much upon it. I will return to possible reasons for its high prevalence in biological sciences later.

What do I mean by 'narrative' in the context of a scientific paper? One way to describe a narrative is that it contains temporality, causality and human interest. It can describe a quest: characters (the investigators) going through a series of events (hypotheses, experiments), one logically, causally leading to the next, and concluding with a climax (discovery). Although the characters, the investigators, are often half-hidden, this will nonetheless sound familiar to anyone having written a paper in the biological sciences recently. The results are not simply listed or tabulated; they are presented as part of a storyline. Typically, a paper will start with an overall setting of context and state the central questions or goals to be explored. In the experimental part, the rationale for doing the first experiment will be provided and the experiment and its results described. These are interpreted and preliminary conclusions are drawn, which lead on to the next experiment, the next element of the story, and so on. Each result is presented as part of a logical flow, directed by the narrator. Scientists even tend to call their paper a 'story'. A paper may be the true story of discovery, but often the reality is somewhat different. Research projects rarely take the shortest or the, in hindsight, most logical path. But in putting the narrative together for publication the presentation is "cleaned up" into a series of logical next steps. The order of events may be changed and rationales may be added after the fact. In addition, some experiments may be inserted at a later time to address requests from reviewers etc. In the end, the results are real, but the narrative path of the paper seldom, if ever, reflects the actual path of the investigation. As in other contexts, a good story pulls the reader along, and ultimately helps them makes sense of, and retain, the new information.

Unlike fictional narratives, scientific papers do not rely on suspense: the summary/abstract states the results and conclusions upfront. Yet, it is interesting that even these short summaries use elements of the narrative form. A recent study investigated this and found a significant positive correlation between narrative elements in the summary and the citation rate of a paper (Hillier et al. 2016). The research area was climate change, which, unlike biology, seems not fully dominated by the use of narrative.

A bit of historical perspective may be useful here. Centuries ago, science was not the highest authority on most subjects. Perhaps in mathematics, but certainly not regarding the physical world. In this context, we can imagine scientists wanting to distinguish scientific discourse from other types of exploration and explanation of the world, be they of religious origin or from other cultural sources. In the natural sciences, the accepted scientific method was, and is, built on careful observations, respect for the facts and objective analysis. The classical format of a scientific paper, dry and factual, reflects this objectivity. Prior to this format's ascendency, the scientific literature was more mixed. Anecdotes and stories occasionally slipped in. But reminders abound to distinguish: "the plural of anecdote is not data" and "just the facts." Simple presentation of method, measurements and calculations should be sufficient for the real scientists, a small and select group back then.

In the twentieth century, Karl Popper outlined a rigorous approach to the scientific method: You cannot prove that a hypothesis is true, but you can prove that it is false by designing experiments to test its predictions. If the results of carefully done experiments are inconsistent with its predictions, the hypothesis must be discarded or modified. Alternatively, if the results meet the predictions, the hypothesis survives by virtue of not being disproven. Often this is done iteratively, with attempts to disprove various competing explanations and hypotheses. Hypotheses gain in strength from repeatedly passing muster, by their predictions continuing to withstand experimental tests with more precise measurements and tests based on new technologies. Eventually the weight of the evidence leads to acceptance. Still, as major and minor paradigms do get overturned or modified over time, the prudence of the distinction between something being proven and not having been falsified (yet) can be appreciated.

While the role of using experiments to challenge hypotheses is generally accepted, presentation of scientific findings seems to have lagged behind. This was pointed out by Peter Medawar, a Nobel-prize winning biologist, in his talk "Is the scientific paper a fraud?" To this provocative title, he answers Yes. Here, he is talking about the classical scientific paper that presents background, methods, a bunch of results and finally the conclusions drawn—in that order. He argues that the structure gives the appearance of deduction: From the results flow the conclusions, as if physical (including biological) sciences were deductive sciences. But hypotheses, ideas and predictions generally come before the experimental results. As they should, according to Popper. So it is false to present the process of scientific discovery the other way around.

Seen in this light, the narrative approach to presenting scientific discoveries seems reasonable and logical, especially for truly hypothesis-driven research. The form can be used to reflect that the investigator starts off not only with known facts, but also with preconceived notions, and that the discoveries are made by designing series of experiments to test ideas in succession, as they develop. Roald Hoffmann, another Nobel laureate and a theoretical chemist, has described narrative as a natural and logical form of scientific communication for both the creating mode (i.e. making new compounds) and for the discovery mode (Hoffmann 2014). He regrets that this form is not valued sufficiently (in the chemistry literature). In biological sciences, papers are generally of the discovery type: observing and interrogating biological entities. Some follow the lines of: "We have this hypothesis and we test it as follows." Others would be better described as: "Look what we found!" But for both, the narrative would be the standard presentation.

Returning for a moment to "just the facts" presentation of science—we must recognize that it is never just the facts. Even before any interpretation, the scientist has chosen which experiments to perform, which data to collect. Naturally, the exploration cannot simply be wishful cherry picking. Attempting to disprove your own hypothesis, not just the competing ones, should instruct the experimental path.

And indeed, the review process does tend to direct an author's attention that way, if it has not been done sufficiently. There are also instructive rules about what is not allowed in science, such as leaving out inconvenient results once obtained.

Publishing in Biological Sciences

The biological sciences make particularly heavy use of narrative. Most papers present relatively long "full stories". How did we get here? I have no definitive answer, but speculate that a few interrelated factors have contributed: The nature of the subject, the size of the field (number of scientists), and the success of certain types of journals.

Biology as a discipline is diverse. The diversity of subject matter—from the macroscopic scale to the molecular scale and from trees to humans—means that there are many niches to work in, each with potential for significant discoveries. As in other fields, the big discoveries, like the mechanisms of evolution or of genetic inheritance, are few and far between. Within the broad subject area, some subfields have grown particularly large. The advances in molecular biology from the 1950s onwards—the structure of DNA, the central dogma (DNA to RNA to protein), the genetic code and the first of many technical advances, restriction enzymes—expanded the scope of this subfield, both in terms of what could be studied and who could do it. Unlike particle physics or astronomy, large and expensive machinery is not essential. Many smaller, non-centralized, laboratories can contribute. Finally, the availability of funding has driven the growth of specific research areas, particularly those of medical relevance. The result is a large and diverse group of scientists in the biological sciences. They are active both as producers and users of data, and as authors and readers of the scientific literature.

Anyone working in the biological sciences is aware of the staggering volume of papers published every year. You simply cannot read everything published in your field. Let's take an example: Cancer is too broad. Apoptosis, then: the study of programmed cell death. In 2017 alone, there were over 26,000 new publications with the word "apoptosis" in the abstract. Each paper will, or should, present a unique set of novel results and their interpretation. Of course not every new result is equally important. But the diversity of subject matter coupled with ever-improving experimental tools means that the massive flow of papers will continue. To help scientists keep up with all this information, the publication process usually involves a ranking of the findings. At the low end are those perceived to have limited importance to a specialized subfield. At the high end are discoveries of perceived high impact and broad interest. This necessary, but somewhat subjective, evaluation of scientific papers has major effects on the dynamics of science. It affects how scientists are evaluated and it empowers journals that publish the highest ranked papers. Communicating effectively to your peers has become a key skill for scientists; it may now be a prerequisite for success. Selecting for good communicators, good storytellers, has, I suspect, also contributed to the prevalence of the narrative form. Others will have noticed what works for the field leaders and will have adopted the successful strategies.

The boom of molecular biology starting in the 1970s was paralleled by a change in publishing. A new journal called Cell became very successful. It championed full papers that presented new and exciting results of general interest, embedded in a well-told story. Then and now, a standard paper might have seven figures, but with each figure usually having multiple parts, the total number of different experiments presented in a paper can be quite large. It requires a well-crafted storyline, one without unnecessary digressions from the plot, for such papers to be easily—and broadly—appreciated. The two other most prestigious journals, Science and Nature, publish shorter papers, but now even these papers tend to be full stories, presented concisely in the main paper and with much of the data provided in the form of supplementary information (online). In the analysis of abstracts mentioned previously, Science and Nature are among those with highest narrative index (Hillier et al. 2016). All three journals aim for "significant novel insights" as well as "general interest". It seems that coupling the appealing writing form of "a good story" with actual novelty and perceived general interest has come to define "top stories". These top stories are what scientists in the biological sciences are chasing.

In the last couple of decades, publishing within biological sciences has expanded further: there are now many more journals, publishing more papers, often with more experiments per paper. There has been a drift toward increasing expectations for when a paper is considered complete, when a story is a full story. If a new molecular mechanism is characterized in flies, getting a paper into a good journal will usually require that the findings are confirmed in a mouse model or in human cells. There is always more that can be done. Top journals sometimes have alternative, less narrative forms as well, such as resource papers, in which large amounts of data that are deemed useful to many scientists are made available with minimal commentary, or methods papers, again with general usefulness as main criterion. The journals may primarily have added these categories in order to capture large numbers of citations, but having them is beneficial, nonetheless.

Am I saying the journals dictate what scientists should do and value? Not directly. But when publication in specific journals is coveted by scientists and rated highly by the people evaluating them (generally other scientists) as well as funding agencies, the requirements of those journals become the standards scientists set themselves. Peer-reviewers and journal editors, in their roles as gatekeepers to publication in specific journals, maintain the need to meet those exact expectations. Many scientists and scientific organizations now lament the power of top journals, even if their own behavior continues to empower them. Breaking away from this cycle is hard, because it is so deeply embedded in scientists' self-evaluation.

Unintended Consequences of Using Narrative

Using a narrative improves readability of scientific papers, especially for readers beyond a narrow set of specialists. It can help us remember the data, and provide a context for understanding. In most cases, it also reflects the actual scientific process better than the dry, classical form. So what might the downside be? I will return later to problems that have to do with scientists and their careers. Of more general concern is the potential impact of thinking in terms of narrative while doing science and of reading scientific results as stories.

The story in a scientific paper is based on data. It is not made up. But, the authors tell the story. Going from an initial set of data to a story has two phases: The first phase is collecting observations and identifying what the story is: in other words, developing the hypothesis. This requires being observant and noticing potentially interesting results, maybe combining them in new ways, and then testing for reproducibility. The second phase is filling out the story: testing the hypothesis as well as doing additional experiments that seem to be required to make the narrative work. This phase is more constrained. Many experiments will be done to rule out artefacts and to exclude alternative (less preferred) hypotheses. Some will be done to test the hypothesis itself. In this phase, scientists typically know which results are more "desired". This creates a bias that may be hard to fully ignore. In acknowledgement of how hard it can be, high quality clinical trials are now always done double-blinded (i.e. neither doctor nor patient knows whether the pill is drug or placebo). A third phase—essentially an extension of the second—can arise when a paper is being considered for publication. Reviewers identify a missing link or an inconsistency in the story—or they think of a logical next step. They ask for specific experiments to be performed to address it. This procedure is intended to improve a paper and it can do so. But now the desired outcome is also even clearer, as the paper will only be accepted if the reviewers are satisfied. The risk of bias, or even misconduct, increases. A situation of this sort was explored in my first novel, Raw Data.

In addition to the impact on what is presented in published papers, there may be a problem of what does not get published when narrative is the norm. There may be a temptation to leave aside certain results because they do not fit the developing narrative, potentially keeping authors and readers on the wrong track. Other observations, even if reproducible and potentially important, may be set aside because they are unrelated to the narrative the scientist is currently working on. If they do not become the focus of another story, such observations tend to remain unpublished and unknown to the rest of the field.

In terms of reading the scientific literature, the elements that help can also constrain and limit critical thinking. A good story is seductive. Simplicity, or Occam's razor, is one of the guiding lights of science and is equally seductive. If something looks logical and straightforward, it is easy to think that the explanation is correct. If the elements of the authors' story line up to give a neat trajectory, it is easy to take it at face value. Had we, as readers, seen the original data in a form that did not provide a ready context, we might have come up with alternative ideas and interpretations. The story in a paper is shaped by the choices made along the way, but also by those made after the fact: how to present data in figures and what to highlight in words. If no choices are made, the story gets lost in an unreadable web of possibilities. In general, authors will discuss alterative interpretations and caveats in the discussion section of a paper. At this point, however, the main story has already been presented, making it harder to rekindle interest in alternative interpretations. In addition, if alternative interpretations are too convincingly presented, the paper may be deemed inconclusive, without a clear story, and not get published—or at least not well. Finally, it is conceivable that the seductiveness of a good story can lead readers, the scientists' peers, to evaluate—and to value—the storytelling as much as the underlying findings.

Scientists, especially younger ones, are completely dependent on publications for their livelihood. They need publications to get a job and to get the grants that allow them to get more work done. Publications mean recognition. However, if publication requires a full story with its many links, it can take a long time. The missing elements of a story may require expertise they cannot easily access—other animal models or analytical methods, for example—and thus involve collaborators who have many other jobs to do. In the meantime, someone else may publish a related finding and the story loses novelty and significance. Publishing quickly in a less demanding journal could be a solution. But in competitive fields, a poorly recognized publication does not help a young scientist much. Being smart and having interesting ideas for what to do in the future is usually not enough. A scientist needs a track record. Young and ambitious scientists will read that as: I need top publications. What this can easily mean is that way too much of a scientist's time and effort ends up being directed at finishing up stories and getting them published, with correspondingly less time and effort left over for innovation and exploring new ideas.

An Alternative: Publishing Single Observations

Experimental scientists know that many important findings arise in the form of a single, illuminating observation. Sometimes it provides the missing piece of a well-known puzzle; sometimes it is unexpected and surprising, opening up new avenues. These days, in our field, such observations would be built up into a narrative for publication. But what if they were simply verified and then published by themselves, without the need to develop a full story around them? I am not trying to deny the usefulness of narrative. But maybe it would be good to have an alternative when publishing science. And preferably one that is also highly respected within the community, to allow for professional development concerns.

One advantage of publishing single observations would be speed. It could be transmitted to the field immediately, years before a full story ever made it to publication. Another advantage could be—for want of a better word—purity: The observation would be unencumbered by the author's narrative. On the other hand, without explanatory context it is possible that people would have a harder time grasping the significance of the finding. Another issue is sheer numbers: having too many observations floating around could make it harder to take proper notice of the relevant ones. This is already a problem with full papers. My suggestion for how these issues could be dealt with is explored in the accompanying novel, Tumble Hitch. It ends up being called Spark publications and is the brainchild of one of the protagonists, Jessie Aitkin. The novel includes a —fictional, of course—description of how it plays out. Real-world initiatives to publish single observations have recently emerged, one of which is called ScienceMatters (www.sciencematters.io). Spark and ScienceMatters both generate single-observation based citable publications, but their approaches are otherwise quite distinct, as discussed below. A different kind of initiative, called nanopublications (nanopub.org), comes from the fields of bioinformatics and text-mining and seeks to deal with both the many large datasets and the many publications generated in biological sciences. Nanopublications are essentially minimal units of author-attributable information that are embedded within narrative papers or within larger datasets deposited to databases, but are rendered machine-readable. Direct mining of nanopublications should allow automated information gathering to be more effective. Overall, it seems fair to point out that these new ideas and initiatives have yet to be endorsed by most active scientists in the field.

ScienceMatters is an initiative from a group of leading life-scientists, supported by academic institutions (i.e. not promoted by a commercial publisher). The short papers/observations are reviewed and published relatively quickly. They must be technically sound and novel. In its present form, ScienceMatters more or less dispenses with the "interest" filter that asks reviewers and editors to evaluate how important a piece of work is, although observations with high scores from reviewers are listed under a separate heading—"Matters Select" as opposed to simply "Matters". Other changes to the traditional mode of publishing include triple blind reviewing to remove personal bias. Thus, like the self-archiving initiative started some years ago, it is less judgmental and more inclusive than the prevailing approach.

ScienceMatters also aims not to rely on the dreaded impact factor. Impact factor measures how much, on average, a paper in a specific journal is cited in the following couple of years. This means it measures popularity in the near term—of both the subfield and of the paper. Despite its many problems, impact factor remains the de facto standard for ranking scientific journals, certainly within biological sciences. In Tumble Hitch, Jessie aims for Spark to be a top-level publication and she cares about its eventual impact factor. One reason is personal: She is an accomplished scientific editor and a high impact factor is her reference point for success. Another is strategic: The aim is to change the publication system from the top. This is attempted by the format, seeking observations of highest interest, and by linking Spark to a top research institute and funding agency as well as to notable figures in the field. With Spark, publication is very fast, as it is based on direct transcripts from small, intense and select meetings. The comments, questions and answers serve as peer-review in this set-up. Comments are directly attributed—so essentially the opposite of fully blinded reviewing—but cannot prevent publication. To be at the top level, selectivity is obviously required. And getting that right has its complications, as touched upon in the novel. At this point, the Spark approach is simply an idea. Maybe it will find its real-life form some day. As an aside, the approach also seeks to enliven and refocus scientific meetings, which for various reasons, including those discussed above, have largely become venues for presenting published work.

Concluding Remarks

Narrative is an ancient and powerful form of communication. In appropriately modified forms, it has numerous useful roles to play in the communication of science. But practicing scientists need to be aware of how they use narrative and how the channeling it imposes may affect their work and their audience. Narrative sets constraints and even traps for the scientist as an author. A well-crafted narrative can seduce the scientist as a reader. A good story is a good story, but it may be worth giving some serious thought to the alternatives as well.

Bibliography and Suggestions for Further Reading

Dahlstrom, M.F.: Using narratives and storytelling to communicate science with nonexpert audiences. Proc Natl Acad Sci USA. **111**(Suppl 4), 13614–13620 (2014)

Hillier, A., Kelly, R.P., Klinger, T.: Narrative style influences citation frequency in climate change science. PLoS One. **11**(12), e0167983 (2016)

Hoffmann, R.: The tensions of scientific storytelling. Am Sci. **102**, 250–253 (2014)

Rørth, P.: Raw data – a novel on life in science. Springer, Dordrecht (2016)

Printed in the United States
By Bookmasters

Printed in the United States
By Bookmasters